STUDIES IN THE
HISTORY OF CHEMISTRY

STUDIES IN THE
HISTORY OF CHEMISTRY

BY

SIR HAROLD HARTLEY, C.H., F.R.S.

HONORARY FELLOW OF BALLIOL COLLEGE, OXFORD

CLARENDON PRESS · OXFORD

1971

Oxford University Press, Ely House, London W. 1
GLASGOW NEW YORK TORONTO MELBOURNE WELLINGTON
CAPE TOWN SALISBURY IBADAN NAIROBI LUSAKA ADDIS ABABA
BOMBAY CALCUTTA MADRAS KARACHI LAHORE DACCA
KUALA LUMPUR SINGAPORE HONG KONG TOKYO

© OXFORD UNIVERSITY PRESS 1971

PRINTED IN GREAT BRITAIN
BY THE CAMELOT
PRESS LTD., LONDON
AND SOUTHAMPTON

TO MY PUPILS

from whom I learnt so much

PREFACE

I OWE my interest in the history of chemistry to my chemistry master at Dulwich, H. B. Baker, 'dry-reaction' Baker, who gave us once a week a fascinating lecture on some classic episode or individual. One day in 1895, when he was looking over my notes, he said to me, 'You are evidently interested in these old men. Why don't you buy their books, you can pick them up very cheaply.' So as a schoolboy I began to spend my pocket-money on them, and as an undergraduate, helped by a legacy from my grandfather, I made a fairly representative collection of the chemical classics in the period 1760–1860 in which I was specially interested. I gave my first lectures on the history of chemistry when I was a demonstrator in the Balliol and Trinity Laboratory in the Lent Term of 1901 in the lecture room in Balliol under the rooms on No. XIX staircase where I had lived as an undergraduate. This staircase has now been replaced by the New Buildings. One evening I met at dinner Charles Cannan, the Secretary to the Delegates of the Oxford University Press. He was always encouraging to young authors and he asked me what I was doing. He said that the Press would like to publish my lectures, and he sent me a contract to sign. However, two of my friends, Henry Miers, the mineralogist, and Edward Poulton, the entomologist, both said that I ought to leave writing history till I was an old man and devote all my time to experimental research. I took their advice and so the book was never written.

However, my interest in the history of science continued and thanks to the books which have been my constant companions I have been able in the leisure of a fairly busy life to write an occasional lecture for the celebration of the centenary of the birth or of some outstanding achievement of a famous chemist. And now, thanks to the kindness of Mr. Colin Roberts, one of Cannan's successors, and of the Delegates of the Oxford University Press, I am able to fulfil my original contract made nearly seventy years ago. It is said to be a record in procrastination. My excuse is that the lectures have probably gained by the delay. They cover much the same ground as my first course and the careful notes I took then, mainly in the library of the British Museum, have often been a help in later years. There are still some gaps I should have liked to fill, notably Henry Cavendish, Wollaston, and Louis Pasteur, but at ninety I mustn't delay any longer and must call it a day. My hope is that this book may encourage the use of history in the teaching of chemistry as these examples of the creative power of the human mind can be so stimulating to the young and the human stories attached to them have their value too.

    I am greatly indebted to the courtesy of the Officers of the Royal Society, the Swedish Academy of Sciences, the Royal Institution, the British Association, the Chemical Society, the Society of Chemical Industry, and the Royal Institute of Chemistry for permission to reprint lectures from their journals.

CONTENTS

List of Plates	xi
1. JOSEPH PRIESTLEY (1733–1804)	1
2. ANTOINE LAURENT LAVOISIER (1743–1794)	19
3. JOHN DALTON (1766–1844) and the Atomic Theory	58
4. HUMPHRY DAVY (1778–1829)	92
5. The Place of JÖNS JAKOB BERZELIUS (1779–1848) in the History of Chemistry	134
6. MICHAEL FARADAY (1791–1867) as a Physical Chemist	153
7. Faraday's Successors and the Theory of Electrolytic Dissociation	173
8. STANISLAO CANNIZZARO (1826–1910) and the First International Chemical Conference at Karlsruhe in 1860	185
9. HENRY ARMSTRONG (1848–1937) and Some of the Great Figures of Nineteenth Century Organic Chemistry	195
10. The Contribution of the College Laboratories to the Oxford School of Chemistry	223
Index	233

LIST OF PLATES

1. Portrait of Priestley by T. Millar — *facing page* xii
2. Portrait of Lavoisier and his wife by David — *facing page* 1
3. *a.* Mme Lavoisier's drawing of a respiratory experiment
   b. Lavoisier and Laplace's ice calorimeter — *facing page* 16
4. Portrait of Dalton by R. R. Faulkner
5. Page from Dalton's note-book
6. Page from Dalton's note-book
7. Portrait of Davy by Sir Thomas Lawrence
8. A page from Davy's note-book — *between pages* 84 *and* 85
9. Faraday's drawing of the safety-lamp
10. Portrait of Berzelius by A. J. Way, 1826
11. Portrait of Faraday by A. Blaikley
12. Title-page of Cannizarro's classic pamphlet — *between pages* 172 *and* 173
13. H.B.H. in 1910
14. Henry Moseley
15. C. N. Hinshelwood's laboratory in 1938 — *between pages* 222 *and* 223

Thanks are due to the following for providing the photographs and for permission to reproduce them: The Royal Society for Pls. 1, 4, 5, 6, 10, and 12; The Royal Institution for Pls. 8 and 9; The Museum of the History of Science, Oxford, for Pls. 13 and 14, and Dr. C. J. Danby for Pl. 15

PLATE 1

JOSEPH PRIESTLEY
From the portrait by T. Millar

PLATE 2

Portraits de M.^r & Mme Lavoisier
d'après le tableau de David.

I

JOSEPH PRIESTLEY
(1733–1804)†

ONE day when Dr. Johnson was being shown some chemical experiments and Priestley's name was mentioned, Johnson knit his brows and said sternly, 'Why do we hear so much of Dr. Priestley?' He was very properly answered, 'Because, Sir, we are indebted to him for these important discoveries,' and Johnson, who disliked Priestley's theological opinions—'They tend to unsettle everything and yet settle nothing'—had to admit the justness of the reply.

We naturally think of Priestley, first as the great pioneer in pneumatic chemistry, whose discoveries played so important a part in the revolution of chemical thought at the end of the eighteenth century. But to do full justice to his genius we must remember that he had no scientific training, that for many years he was a minister and a schoolmaster, that theology, to quote his own words, was 'his original and proper province', and that he was one of the most versatile and prolific of authors. He wrote over one hundred books, and their subjects range over theology, politics, philosophy, oratory, history, and grammar, as well as natural science, and in each of them he won recognition.

He played an important part in the history of philosophic radicalism: it was to him that Bentham owed his famous phrase 'the greatest happiness of the greatest number', and Priestley was the first to associate the 'greatest happiness' principle with the idea of democracy. Throughout his life he was a fearless champion of the causes of religious freedom and

† Extracts from the two Addresses given at the meetings of the Royal Society on 15 March 1933 and at the Chemical Society on 6 April 1933, to celebrate the bicentenary of Priestley's birth on 13 March 1733.

representative Government and with his friends he prepared the way for the great reforms of the nineteenth century.

But in addition to all this, Priestley's unbounded curiosity and his genius for observation and experiments led him to the discovery of most of the common gases, thus opening up a new field to chemists. It is these discoveries that I propose to describe, and I will do so far as possible in Priestley's own words.

He was born near Leeds in 1733, his father being a cloth-dresser who belonged to the Congregationalist body. He was educated for the ministry, and at the age of nineteen he had studied no less than nine languages. After spending three years at a theological academy at Daventry he was appointed minister to a small congregation at Needham Market at the age of twenty-two. His flock soon deserted him owing to his heterodox views and his stammering delivery, and in 1758 he went to a new congregation at Nantwich where he started a school for boys and girls, 'which', he says, 'soon enabled me to purchase some philosophical instruments, as a small air pump, an electrical machine. . . . I had no leisure, however, to make any original experiments until many years after this time.'

The turning point of Priestley's life came in 1761 when he went as tutor in languages and *belles lettres* to the new dissenting academy at Warrington. Whilst teaching there he had the good fortune to meet Benjamin Franklin, who was in London trying to secure the repeal of the Stamp Act. Priestley, finding his other literary occupations nearly ended, proposed to Franklin with characteristic courage to write a history of electricity, if Franklin would lend him the necessary books. Within twelve months the proofs were in Franklin's hands and the book was not a mere recital of other people's work as Priestley added a number of observations of his own. It was published in 1767 and contains the first suggestion that electrical forces obey the law of inverse squares: Priestley based this deduction on the behaviour of pith balls inside a charged cup, which proved, he thought, the analogy between electrical and gravitational forces.

His most remarkable electrical experiment was made on what he called the lateral explosion accompanying a discharge of Leyden jars. Under certain circumstances he showed that on discharge a spark would pass between the discharging rod and an insulated conductor nearby, and as the conductor remained uncharged Priestley concluded that the discharge must be of an oscillatory character. To use his own words, 'an electric spark must enter and pass out again, within so short a space of time, as not to be distinguished, and leave no sensible effect whatever'. He thus anticipated Sir Oliver Lodge's researches on the 'Side-Flash' made in 1888.

In 1767, at the age of thirty-four, Priestley moved to Leeds to take up a new appointment at the Mill Hill Chapel and there a fortunate accident turned his thoughts to chemistry. 'Living for the first year in Leeds in a house that was contiguous to a large common brewery,' he says, 'so good an opportunity produced in me an inclination to make experiments in the fixed air that was continually produced in it. Had it not been for this circumstance I should probably not have attended to the subject of air at all.' Priestley amused himself by making experiments with the carbon dioxide, or fixed air as Black had called it, which was produced in the fermentation vats, and happening to read that the spring waters of Spa are charged with fixed air, 'one of the first things I did in this brewery', he says, 'was to place shallow vessels of water within the region of fixed air; and having left them all night, I generally found the next morning that the water had acquired a very sensible and pleasant impregnation; and it was with peculiar pleasure that I first drank of this water which, I believe, was the first of its kind that had ever been tasted by man.' Some years later Priestley was dining with the Duke of Northumberland and a bottle of water was produced which had been distilled for the use of the navy. 'This water', says Priestley, 'was perfectly sweet, but, like all distilled water, wanted the briskness and spirit of fresh spring water; when it immediately occurred to me that I could easily mend that water for the use of the navy (and perhaps supply them with a cheap and easy method of preventing or curing the sea scurvy), viz., by impregnating it with fixed air.'

And so the first practical result of Priestley's chemical investigations was the invention of soda water. 'A service', says Huxley, 'to naturally, and still more to artificially, thirsty souls, which those whose parched throats and hot heads are cooled by morning draughts of the beverage, cannot too gratefully acknowledge.'

His experiments with carbon dioxide had aroused Priestley's curiosity about other gases, and the first mention of his new researches is in a letter dated 21 February 1770, when he wrote, 'I am now taking up some of Dr. Hales's enquiries concerning air'. Priestley had been reading Stephen Hales's *Vegetable Staticks* published in 1727, and to appreciate the full implication of this we must look back a century to the time of Samuel Pepys when the methods of manipulating gases were first being evolved by Robert Boyle and Hooke.

Pepys was admitted a Fellow of the Royal Society on 15 February 1665, when he wrote in his diary:

Thence with Creed to Gresham College, where I had been by Mr. Povy the last week proposed to be admitted a member; and was this day admitted, by signing a book and being taken by the hand by the President, my Lord Brunkard, and some words of admittance said to me. But it is a most acceptable thing to hear their discourse, and see their experiments; which were this day upon the nature of fire, and how it goes out in a place where the ayre is not free, and sooner out where the ayre is exhausted, which they showed by an engine on purpose. After this being done, they to the Crowne Taverne, behind the 'Change, and there my Lord and most of the company to a club supper; Sir P. Neale, Sir R. Murrey, Dr. Clerke, Dr. Whistler, Dr. Goddard, and others of most eminent worth. Above all, Mr. Boyle to-day was at the meeting, and above him Mr. Hooke, who is the most, and promises the least, of any man in the world that ever I saw.

A century later Priestley's earliest chemical researches were directed to exactly the same problem as the experiments of Boyle and Hooke seen by Pepys—the cause of the extinction of a flame burning in a closed space. A century had passed but chemists were no nearer to an explanation because with rare exceptions they had paid but scanty attention to the existence and behaviour of gases. Mayow, who was elected a Fellow of the Royal Society in 1678, was one of the exceptions—he doubtless had learned much from Boyle and Hooke, but he was an ingenious experimenter himself, and in his collected essays, published in 1674, one of the plates illustrates the methods he used for the manipulation of gases which have survived, with small modifications, to the present day. He anticipated the principle of the McLeod gauge in an apparatus he devised for testing the elasticity of gases in order to decide whether they are true air or not. Luckily for chemistry, Mayow's *Essays* came into the hands of Hales, who developed his experimental methods and through Hales the technique of Boyle, Hooke, and Mayow passed to Black, Lavoisier, and Priestley.

I have disgressed a little in order to show Priestley's place in the great tradition of pneumatic chemistry which has flourished for two and a half centuries—a tradition that has been handed on by Boyle, Hooke, Mayow, Hales, Black, Cavendish, Priestley, Davy to Rayleigh, Ramsay, and Travers in our own time. With it is associated the discovery of nearly all the gaseous elements.

Priestley began by studying the phenomena of combustion and respiration in confined volumes of air and observed the contraction in volume and the vitiation of air. Although he had no idea of the nature of the changes he was studying, he saw at once that there must be in nature some means for maintaining the purity of the atmosphere. To quote his own words: 'The quantity of air which even a small flame requires to keep it burning is prodigious. It is generally said that an ordinary candle consumes, as it is called, about a gallon in a minute. Considering this amazing consumption of air by fires of all kinds, volcanoes, etc., it becomes a great object of philosophical enquiry, to ascertain what change is made in the constitution of air by flame, and to discover what provision there is in nature for remedying the injury which the atmosphere receives by this means.'

This he set out to investigate, his guide being 'generally to consider the influence to which the atmosphere is, in fact, exposed'. He tried the effect of washing foul air with water, of exposing it to light, of heating it and cooling it, and of subjecting it to the action of those 'effluvia which are continually exhaling into the air'. He tried the fumes of burning sulphur and of the smoking spirit of nitre. There he let the discovery of oxygen slip through his fingers for the first time.

How Priestley discovered the effect of vegetation on the composition of the atmosphere is best told in his own words: 'One might have imagined that, since common air is necessary to vegetable, as well as to animal life, both plants and animals had affected it in the same manner; and I own I had that expectation, when I first put a sprig of mint into a glass jar inverted over a vessel of water: but when it had continued growing there for some months, I found that the air would neither extinguish a candle, nor was it at all inconvenient to a mouse, which I put into it.' This unexpected result led him to try the effect of vegetation on different kinds of foul air and he found that in some cases the sprigs of mint he used died quickly, 'but if they do not die presently, they thrive in a most surprising manner. In no other circumstances have I ever seen vegetation so vigorous as in this kind of air, which is immediately fatal to animal life.... This observation led me to conclude that plants, instead of affecting the air in the same way as animal respiration, reverse the effects of breathing and tend to keep the atmosphere sweet and wholesome, when it is become noxious, in consequence of animals either living and breathing, or dying and putrefying in it.' The critical experiment was made in the beginning of August, 1771. 'I took a quantity of air made thoroughly noxious by mice breathing and dying in it, and divided it into two parts; one of which I put into a phial immersed in water, and to the other (contained in a glass jar standing in water) I put a sprig of mint.... After eight or nine days I found that a mouse lived perfectly well in that part of the air, in which the sprig of mint had grown, but died the moment it was put into the other part of the same original quantity of air, and which I had kept in the very same exposure, but without any plant growing in it.' A little later in the same month he showed that a growing plant has also the same power of restoring air in which a candle had burnt out.

Thus within a year Priestley had solved his problem in spite of the handicap that he had no idea of the nature of the chemical changes involved. His immediate recognition of the wider issues raised by his experiments and his speedy solution of the problem are a striking example of his vision and of the instinct that guided his investigations.

Like his contemporaries, Priestley thought that substances are inflammable because they contain a mysterious principle, called by Stahl phlogiston, which is liberated together with heat and light so long as there is a supply of air to remove it. Respiration was thought to be a similar process, the air removing phlogiston from the lungs. This theory was a relic of alchemy, and in order to explain the observed facts it had to assume that phlogiston could, when necessary, confer the properties of inflammability, levity, metallic lustre, colour, elasticity, volatility, and smell. In fact, arguing with a phlogistic chemist must have been like playing cards against a hand of jokers. This looking-glass hypothesis survived for almost a century, until Lavoisier exposed its fallacies by his study of the changes of weight during combustion, viewed in the light of Priestley's experiments on gases.

In his early experiments the only test Priestley had of the 'goodness of air' was to see how long mice could live in it, but in 1772 a new discovery gave a convenient and accurate means of measuring this property. Mayow, by the action of iron on spirit of nitre (nitric acid), had obtained a colourless gas (nitric oxide) which gave red fumes if mixed with air, the red fumes being soluble in water. Hales had obtained the same gas and showed its action on air by dissolving Walton Pyrites in spirit of nitre, and Priestley wishing to repeat his experiments asked Cavendish's advice as to the possibility of obtaining the same mineral as Hales had used. Cavendish early in 1772 advised him to try other kinds of pyrites and metals, suggesting that the essential factor was the spirit of nitre. 'I presently', says Priestley, 'found what I wanted, and a good deal more.' He obtained the gas to which he gave the name 'nitrous air' by the action of spirit of nitre on a number of different metals, and he found that when mixed with common air over water it gave red fumes which were absorbed by the water with a considerable diminution of volume amounting to about one-fifth of the common air, and as much of the nitrous air as was necessary to produce this effect, i.e. about half the volume of air taken.

Priestley decided that the 'diminution, occasioned by the mixture of nitrous air, is peculiar to common air, or *air fit for respiration*, and is at least very nearly, if not exactly, in proportion to its fitness for that purpose; so that by this means the goodness of air may be distinguished much more accurately than it can be done by putting mice, or any other animals, to breathe in it. This was a most agreeable discovery to me, especially as from this time I had no occasion for so large a stock of mice.' From this experiment of Priestley's has grown the science of gas-analysis, and the word 'eudiometer', literally the measurer of good weather, reminds us of the purpose for which it was first used.

It occurred to Priestley in 1772 that by using mercury he could collect gases that were soluble in water, and by this means he was able to isolate marine acid air (hydrogen chloride) which afforded a striking example of the varied properties which gases may possess.

In 1772, Priestley, who was growing restless at Leeds, gladly accepted an invitation from Lord Shelburne to become his librarian. His new duties were almost nominal, leaving him ample leisure for his experiments and writing, and the next seven years were rich in new discoveries; the first of these was ammonia, which Priestley describes in a letter to Franklin dated 26 September 1773: 'The most important of the observations I have lately made is of an alkaline corresponding to the acid air, of which an account is given in what I have already printed. This I get by treating a volatile alkali in the same manner in which I before treated the spirit of salt. As soon as the liquor begins to boil, the vapour arises, and being received in a vessel filled with quicksilver, continues in the form of air well condensed by cold. I imagined that a mixture of this alkaline with acid air would make a neutral and perhaps a common air, but instead of that they make a very beautiful white salt of a very curious nature. Nitrous air makes this alkaline vapour turbid, and perhaps generates a different salt. But I have not yet made a tenth part of the experiments that I propose to do with this new kind of air. . . . If volatile alkali, liquid or solid, be exposed to nitrous air during its effervescence with common air, the vessel is presently filled with beautiful white clouds, and the salt is tinged blue, this explains the constitution of nitrous air, but I have no time for reasoning.' Thus Priestley makes it clear that it is to his experiments and not to his deductions that he attaches importance.

Early in 1774, Priestley discovered nitrous oxide by the action of moist iron and sulphur on nitrous air (nitric oxide). He showed that the new gas is much more soluble in water and he was surprised to find that a candle burnt in it with an enlarged flame. He was doubtful as to its nature: the evidence of the candle flame indicated that it was dephlogisticated nitrous air, but as it was formed by the action of iron and sulphur it might equally well have been phlogisticated nitrous air. Priestley afterwards obtained it by the action of zinc and tin on dilute nitric acid. It was with this gas that Davy, twenty-four years later, discovered the first anaesthetic. One of those to whom Davy administered the gas was the poet Southey and from the account of his sensations which he gave to his brother we see why it received the name of 'laughing gas'. 'Oh, Tom', he wrote, 'such a gas has Davy discovered, the gaseous oxyde. Oh, Tom, I have had some; it made me laugh and tingle in every toe and finger tip—Davy has actually invented a new pleasure for which language has no name. Oh, Tom, I am going for more this evening; it makes one strong and so happy, and without any after debility, but instead of it, increased strength of mind and body. Oh excellent air-bag. Tom, I am sure the air in heaven must be this wonder-working gas of delight.'

The next new gas which Priestley isolated was sulphur dioxide, or, as he called it, vitriolic acid air. He got it by the action of hot vitriolic acid on substances rich in phlogiston, e.g. various metals, charcoal and animal fats.

On 1 August 1774, Priestley obtained oxygen by heating mercuric oxide. It was not the first time that this gas had been in his hands, but on that occasion it did not escape his notice, although several months elapsed before he realized its real nature. The course of these experiments gives a good idea of the way in which Priestley worked.

In 1774 he was given a large burning glass, and he used it in order to heat a number of bodies confined over mercury in glass tubes. Among the substances he took were the calces (or oxides) of mercury and lead, and he noticed that both of them gave off considerable quantities of gas in which a candle burnt with an enlarged flame as in altered nitrous air. He concluded that he had obtained altered nitrous air although the manner of its production puzzled him greatly. He satisfied himself that no nitrous acid had been used in preparing the calx of mercury, but he thought that possibly, during its preparation, the calx had collected 'something of nitre, in that state of heat, from the atmosphere'. A few days later he started on a continental tour with Lord Shelburne, and when they were in Paris in the following October he told Lavoisier and other French chemists of his latest discovery. 'At the same time I had no suspicion that the air which I had got from the mercurius calcinatus was even wholesome, so far was I from knowing what it was that I had really found; taking it for granted that it was nothing more than such kind of air as I had brought nitrous air to be by the process above mentioned; and in this air I have observed a candle would burn sometimes quite naturally, and sometimes with a beautifully enlarged flame, and yet remain perfectly noxious.' On returning to England he continued to investigate the gas, and finding that, unlike phlogisticated nitrous air, it was hardly soluble in water, he concluded it must be a different substance. 'In ignorance of the real nature of this kind of air, I continued from this time (November) to the 1st of March following . . . till then I had so little suspicion of the air from mercurius calcinatus being wholesome that I had not even thought of applying to it the test of nitrous air.' The first time he applied the nitrous air test he took one measure of nitrous air to two measures of the new gas, and as, owing to the quantities he took, he found a diminution only slightly greater than with common air, he concluded that the air was, in fact, common air . . . 'this advance in the way of truth, in reality, threw me back into error, making me give up the hypothesis I had first formed, viz., that the mercuric calcinatus had extracted spirit of nitre from the air; for I now concluded, that all the constituent parts of the air were equally, and in their proper proportion, imbibed in the preparation of this substance, and also in the process of making red lead.' Then the next morning he tried the residue from the experiment in nitrous air, and while it ought to have been noxious it allowed a candle to burn better than in common air. 'I cannot, at this distance of time, recollect what it was that I had in view in making this experiment; but I know I had no expectation of the real issue of it. Having acquired a considerable degree of readiness in making experiments of this kind, a very slight and evanescent motive would be sufficient to induce me to do it. If, however, I had not happened, for some other purpose, to have had a lighted candle before me, I should probably not have made the trial.' He still thought this might be accidental and the air was much the same as common air. 'I wanted a mouse to make the proof quite complete. On the eighth of this month I procured a mouse, and put it into a glass vessel, containing two ounce-measures of the air from mercurius calcinatus.' It should have lived a quarter of an hour and actually lived half an hour. Still, he did not conclude the air was better, 'so little accuracy is there in this method of ascertaining the goodness of air'. Next day he tried the residue left by the mouse with nitrous air and found it better than common air. 'Thinking of this extraordinary fact on my pillow, the next morning I put another measure of nitrous air to the same mixture, and, to my utter astonishment, found that it was farther diminished . . .' He then tried more mice and nitrous air tests until 'being fully satisfied of the superior goodness of this kind of air, I proceeded to measure that degree of purity, with as much accuracy as I could.' He found it between four and five times as good as common air.

Priestley was now convinced that he had obtained a new species of air which he named dephlogisticated air, because it could take up more phlogiston than ordinary air. Lavoisier, on the other hand, was quick to realize that the new gas was the final clue he needed to explain the problem of combustion. He named it oxygen and made it the central element of his new system, which marks the birth of modern chemistry. But while Priestley failed completely to grasp the inner significance of his discovery, he saw at once the possibilities of its practical applications. 'Nothing', he writes, 'would be easier than to augment the force of fire to a prodigious degree, by blowing it with dephlogisticated air instead of common air. . . . Possibly much greater things might be effected by chemists, in a variety of respects, with the prodigious heat which this air may be the means of affording them.' And again: 'From the greater strength and vivacity of the flame of a candle, in this pure air, it may be conjectured, that it might be peculiarly salutary to the lungs in certain morbid cases, when the common air would not be sufficient to carry off the phlogistic putrid effluvium fast enough. But perhaps we may also infer from these experiments, that though pure dephlogisticated air ought to be very useful as a *medicine*, it might not be so proper for us in the usual healthy state of the body: for, as a candle burns out much faster in dephlogisticated than in common air, so we might, as may be said, *live out too fast*, and the animal power be too soon exhausted in this pure kind of air. A moralist, at least, may say, that the air which nature has provided for us is as good as we deserve. My reader will not wonder that after having ascertained the superior goodness of dephlogisticated air . . . I should have the curiosity to taste it myself. I have gratified that curiosity by breathing it . . . The feeling of it to my lungs was not sensibly different from that of common air, but I fancied that my breast felt peculiarly light and easy for some time afterwards. Who can tell but that, in time, this pure air may become a fasionable article in luxury. Hitherto only two mice and myself have had the privilege of breathing it.'

Never was an inventor's vision better justified: each year in Europe and America some 5 000 000 000 ft^3 of oxygen are used, as Priestley had foreseen in industry and medicine.

After the discovery of oxygen Priestley continued his investigations over the whole range of pneumatic chemistry, without any systematic objective, but adding continuously to our knowledge of gases. His next discoveries were silicon fluoride, chamber crystals, and nitrogen dioxide, the first gas to be collected by downward displacement. He was intrigued by its change of colour on heating, and carried a bulb about to show this to his friends. He found that ammonia was formed from nitric acid by what we should call today an iron–copper couple, and that it was decomposed both by heat and by the electric spark yielding inflammable air.

The certainty with which he recognized again any gas he had previously examined was remarkable, and he had a flair for finding out the simplest means of preparing them. For instance, nitric oxide he made from copper and nitric acid, and nitrogen dioxide by heating lead nitrate—both are the methods that we use to-day.

He compared the decomposition of the salts of hydrochloric, nitric, and sulphuric acids, and sagaciously remarked that the first differs from the others as its salts never yield any dephlogisticated air, i.e. oxygen.

He investigated afresh the melioration of air by the growth of plants and showed that light was an essential factor in the purification. He also studied the respiration of fish and found that their existence depends on air dissolved in water.

Another aspect of Priestley's work reveals him as one of the earliest physical chemists.

Following Cavendish, he determined roughly the relative densities of gases by weighing balloons filled with them. He also examined the volume ratios in which a number of pairs of gases combine, but although these were nearly always in the neighbourhood of simple numbers, he did not speculate as to the significance of this. He observed the relative solubilities of gases in water and used these as a means of identifying them.

Priestley measured the expansion of gases by heat in a crude apparatus, but the coefficients he found were not, alas, identical for different gases. However, he noticed that ammonia appeared to have a much greater coefficient than the rest, and he ascribed this rightly to the presence of water in his apparatus and the smaller solubility of ammonia at higher temperatures. He also compared the conductivity of sound in different gases by means of a clockwork bell in a receiver which he filled with them. His conclusion was that 'the intensity of sound depends solely on the density of the air in which it is made, and not at all upon any chemical principle in its constitution'.

He then compared the conductivity of heat in various gases by means of a thermometer in a glass bulb which he put into hot and cold water and observed the rate of change of temperature when the bulb contained different gases. He found that inflammable air (hydrogen) conducted heat better than any other gas and that fixed air is a worse conductor than common air.

Priestley was one of the first photo-chemists, as he discovered that sunlight was necessary for the purification of air by a growing plant and he showed the effect of light in decomposing nitric acid. 'Light', he said, 'besides serving the important purpose of vision is likewise a chemical principle, the effects of which are as yet but little known.'

He made use of the electric spark both for making gases enter into combination, e.g. hydrogen and oxygen, and for decomposing them, e.g. ammonia. He tried to compare the electrical conductivity of different substances, and he was one of the first to describe the phenomena accompanying the electrical discharge through gases at low pressures.

'The application of electricity to chymistry', he wrote, 'is as yet in its infancy. I have made some use of it in the doctrine of air, and Mr. Cavendish has made a most capital one. This should encourage us to apply it with more assiduity.'

Having determined the relative densities of different gases Priestley began to speculate as to their behaviour when mixed, and this led him to the first investigation of the problem of gaseous diffusion. 'Considering the very different specific gravities, and other remarkedly different properties of different kinds of air,' he says, 'it might naturally enough be taken for granted that those which differ very much in specific gravity at least, would separate from such other after they were mixed.' This conclusion he proceeded to examine by mixing fixed and common air, inflammable and nitrous air, nitrous and common air, nitrous and fixed air, and leaving them to stand in a cylinder. He then expelled portions of the mixtures very carefully by admitting water at the bottom of the cylinder, and tested the composition of the early and later fractions. 'The result of my trials has been this general conclusion; that when two kinds of air have been mixed, it is not possible to separate them by any method of decanting them or pouring them off . . . They may not properly incorporate, so as to form a new species of air possessed of new properties; but they will remain equally diffused through the mass of each other.' Thus Priestley established the irreversibility of gaseous diffusion.

In spite of the crude methods by which these physical measurements were made, and the inaccuracy of the results, taken as a whole they are a remarkable achievement, especially when we consider how limited was Priestley's general scientific experience. They illustrate once again his courage and initiative as an experimenter. Undoubtedly, too, the evidence that the various gases possessed definite but different physical properties must have helped to establish their individuality in the minds of chemists.

In 1780, Priestley parted with Lord Shelburne and went to Birmingham. There was no quarrel between them and they seem to have continued on very friendly terms. Talleyrand, in his *Memoirs*, says that he found Priestley at Lansdowne House after the destruction of his home in 1791.

In Birmingham he found a very congenial circle of friends, including Boulton and Watt, the engineers, Josiah Wedgwood the potter, Samuel Galton, and Erasmus Darwin, and he soon became minister of the New Meeting. He set up his instruments in an outhouse which he called his 'elaboratory' and continued his experiments. He examined the influence of gases on the conductivity of heat and sound and was the first to show that both vary with the density of the gas, the lighter gases conducting heat more rapidly but being poor conductors of sound. Apart from this there were few discoveries, and much of Priestley's energy was directed to fighting the new views of Lavoisier which were now rapidly gaining ground. He showed great ingenuity in finding difficulties which the new theory could not explain, the most serious of them being due to the confusion that existed between carbonic oxide and hydrogen, owing to the fact that both are inflammable gases.

But he was also engaged in other controversies that were to have more serious results for him. In 1789 he preached a sermon in support of the attempt that was being made to repeal the Test Act. This was violently attacked from the Episcopal pulpits and he was drawn into a bitter quarrel. 'To dispute with the Doctor was deemed the road to preferment. He had already made two Bishops and there were several heads that wanted mitres.'

After the French Revolution popular feeling was aroused against the party of Reform in this country, and particularly against Priestley who had written a pamphlet vindicating the principles of the Revolution. The cry of 'Church and King' was raised against the Liberal Dissenters, and on the occasion of a dinner in Birmingham in 1791 to celebrate the second anniversary of the taking of the Bastille, a mob collected and after wrecking Priestley's chapel, went to his home, which was burned with all his instruments and papers. Fortunately he had been persuaded to leave shortly before their arrival, and he escaped to London.

Priestley accepted this cruel stroke of fortune with equanimity and patience, and within a few months he had settled down at Hackney to a new congregation and was teaching at the Hackney New College. However, finding himself shunned for his political and religious opinions, and being mistrustful of the future at home, in 1794 he followed his sons who had emigrated to America. There he lived on the banks of the

Susquehanna, continuing his theological and scientific studies almost until the day of his death in 1804.

Although in public life Priestley was such a keen controversialist, disregarding all personal considerations when he was fighting for the causes which he held so dear, in private life he was one of the most gentle and kindly of men, beloved of children; even his opponents could not resist the charm of his personality. No man had more devoted friends, and in the crises of his life they vied with one another in giving him proofs of their affection. Samuel Galton's daughter described him as 'the father of discoveries on air, a man of admirable simplicity, gentleness and kindness of heart, united with a great acuteness of intellect. I can never forget the impression produced on me by the serene expression of his countenance.'

No one knew better than Priestley where his strength lay in science. Just before his death he wrote to Davy, 'I was a discoverer before I was a chemist', and twenty years earlier he said, 'Upon this as on other occasions, I can only repeat that it is not on my opinions on which I would be understood to lay stress. Let the new facts from which I deduce them be considered as my discoveries and let other persons draw better inferences from them if they can.' Compare this with a sentence of Lavoisier's dealing with the relationship of his own work to Priestley's: 'As the same facts have led us to diametrically opposite conclusions, I hope that, even if I am reproached with having borrowed my evidence from the work of that distinguished philosopher, no one will deny that the deductions I have made are my own.'

Priestley was in competition with men of the calibre of Black and Cavendish without their training and knowledge and without their resources, and yet he made more discoveries about the common gases than anyone else. Black and Cavendish worked more intensively on certain gases, but Priestley's great service to chemistry was the *extensiveness* of his work which revealed to chemists the variety of substances which could exist in the gaseous state, their individuality, and the importance of the part they played in chemical reactions. No longer could chemists believe that there was 'only one kind of air loaded with different vapours'.

What was the secret of Priestley's genius as an investigator?

PLATE 3

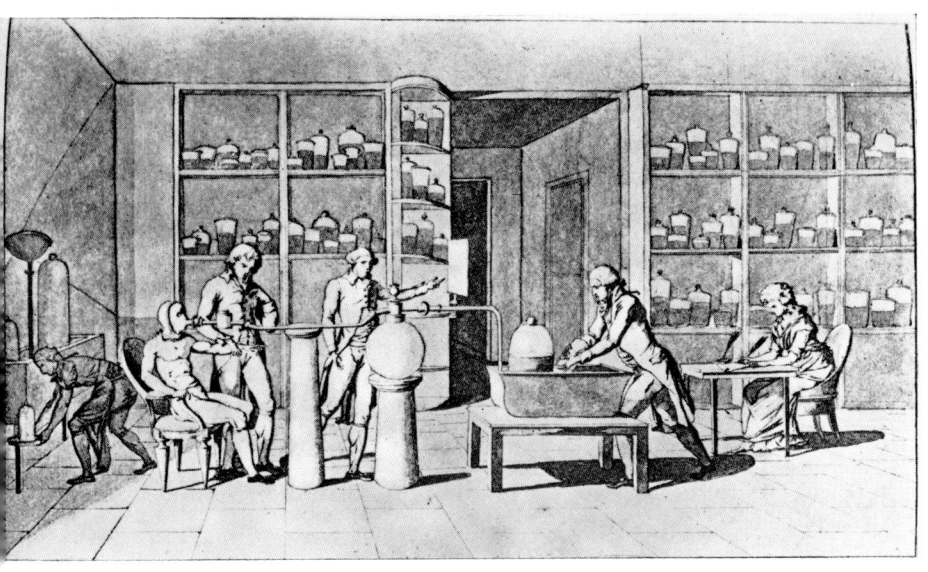

a. Mme Lavoisier's drawing of a respiratory experiment

b. Lavoisier and Laplace's ice calorimeter

His enthusiastic energy and curiosity, his fertility of mind, and the ingenuity and enterprise with which he devised and carried out new experiments with the simplest means; his keen observation, his exceptional visual memory and the diligence and rapidity with which he developed each investigation. And behind everything there was the impelling force of Priestley's love for science, his devotion to experiment, and his boundless faith in the possibilities of new scientific discoveries. This is a passage from a letter to Lord Shelburne in 1778:

My view in advising your Lordship to establish and furnish a laboratory for philosophical purposes was double: First, to accustom Lord Fitzmaurice, at an early age, to the use of philosophical instruments, and the sight of philosophical experiments and processes, in order to do for him, if he should happen to acquire a taste for natural science, what all his fortune will not otherwise be able to do, viz., to make him happy in active and pleasing pursuits at home; and I know of nothing in the range of human life that can answer this invaluable purpose so well. Mere literary pursuits are generally unfavourable to health or cheerfulness, though they may contribute to amuse and tranquilize the mind.

My other view was to prosecute original inquiries under your Lordship's auspicies, to indulge my own inclination and ardour in these pursuits, and at the same time to make myself really useful to your Lordship's general fame and character.

After a reference to his new volume of experiments in which he wishes to acknowledge his debt to Lord Shelburne, he concludes:

I am the more desirous of doing this, as by some of my other publications I may involve your Lordship in some part of the odium I bring upon myself with the ignorant and narrow-minded. But wherever I go, I must be taken for better and for worse; and I think it the first of duties to that Being who has given me whatever faculties I am possessed of, to pursue and propagate at any risk, all important truth.

This is one of the most revealing passages in all Priestley's writings: 'I must be taken for better and for worse . . . to pursue and propagate at any risk, all important truth.' We see him there with his frank simplicity and directness of purpose, incapable of compromise, and regardless of personal considerations when principles were at stake.

Let me end with a phrase from Dante, which seems to me to describe so well Priestley's eager restless spirit, fearless and untiring in the defence of liberty and in the quest of truth—
con l'ali snelle . . . del gran disio
Borne on the swiftly beating wings of great desire.

2

ANTOINE LAURENT LAVOISIER
(1743–1794)†

ANTOINE LAURENT LAVOISIER is one of the immortals. In the whole history of science there is no transformation so swift and dramatic as when in 1789, in his great Treatise, he gave chemistry its modern form, sweeping away the cobwebs of centuries which obscured its progress. He was a child of his age. Bred in France at a time when the writings of Voltaire and Rousseau were stirring men's minds, and Diderot and the Encyclopaedists were widening their vision, Lavoisier was a great reformer. It is difficult within the compass of an hour to do justice to his many-track mind. Ambition, curiosity, humanity, and love of action took him into many fields, and in each his creative genius saw an opportunity for constructive work. The six great volumes of his collected writings display an amazing intellectual and practical achievement. In chemistry, in physics, in physiology, in chemical engineering, in agriculture, in geology, in education, in statistics, and in finance he was a pioneer, and in each his contribution has the modern touch both in thought and phrase. The marvel is that with his widespread interests it was he alone, in a band of brilliant contemporaries like Black, Cavendish, Priestley, and Scheele, who had the vision of modern chemistry and gave it life. Alone among the intellectuals Lavoisier was a man of affairs, skilled in administration and finance, and that is in some measure the secret of his success. For it can be said of him that he applied not only the balance but the principle of the balance sheet to chemistry and physiology.

Lavoisier's life was such a tangled skein of occupations that they cannot be seen in their true perspective under separate headings. Their reactions on each other are the clue to so much that, even at the sacrifice of logical sequence, I shall try

† Memorial Lecture at the Royal Society on 16 November 1944.

to picture the daily current of his life flowing in its many channels.

Lavoisier was a Foreign Member of the Royal Society and, although he never came to England, at many critical moments of his scientific work he was influenced directly or indirectly by Fellows of our Society. You will see that Robert Boyle, John Mayow, John Locke through his disciple Condillac, Stephen Hales, that most ingenious vicar of Teddington, Black, Priestley, and Cavendish all had their influence on Lavoisier.

EARLY LIFE

He was born in Paris on 26 August 1743. His father was a lawyer, and his mother and grandmother came of legal stock, so that he had the law in his blood, and no one could draft an opinion more skilfully. After a classical education at the Mazarin College, he turned to science, and was fortunate in his teachers. Mathematics and astronomy he learnt from La Caille, chemistry from Rouelle, geology from Guettard, and botany from de Jussieu. From their teaching not only did he gain a broad general scientific background, but he was bitten with the idea of research.

He qualified as a lawyer, but he could afford to choose his own career, and his ambition turned to science. There was a touch of Francis Bacon in his make-up, a belief that by making systematic measurements the truth would emerge. He began like Dalton as a meteorologist, and his first published paper was on an aurora.[1] Throughout his life he kept careful records of daily thermometric and barometric observations, made by himself and by correspondents in different parts of the world. He was always meaning to collate them, but that was never done, and his only meteorological paper was on forecasting, in which he suggested the value of a daily weather forecast.[2] It was in meteorology that he served his apprenticeship in physical measurement, investigating the accuracy of the barometer and thermometer, and the best construction of the portable instruments which he sent to his correspondents.

GEOLOGY

In 1763, when he was twenty, Guettard interested him in his project of a geological map of France, the first of its kind. It appealed to Lavoisier's systematic mind, and he threw himself into the scheme with characteristic energy, writing to ministers for support, and drawing up questionnaires to go to the provinces. For seven years Lavoisier was making geological tours to collect data for the survey, and his journeys gave him a cross-section of social conditions all over France that left its mark on his mind.

He travelled on horseback with his servant carrying his barometer, thermometer, hydrometer, and box of reagents. His observations included both mineral deposits and the thickness and characteristics of the various strata at different exposures. He measured in each case their height above sea-level with his barometer, so that he knew the levels at which they occurred in different places. His material formed the basis of sixteen plates in the first geological map of France which was published in 1784 by Guettard's successor Monnet without proper acknowledgement of Lavoisier's work. His only substantial geological paper, on recent sedimentary rocks,[3] was not published until 1788. In it he distinguishes between littoral and pelagic banks which were formed at different distances from the land and consisted of distinct kinds of sediments and organisms. His sections gave the first outline of a correct classification of the Tertiary deposits of the Paris region. Geikie said of Lavoisier that 'if he had not given himself up to chemistry, he might have become one of the most illustrious among the founders of geology'.

PUBLIC LIGHTING

Throughout his life Lavoisier's interests were divided between pure science and its application to practical problems. He had a deep sense of responsibility that science should be used in the service of man, and he was always ready to turn aside from his researches to some practical task. In 1765 the Academy offered a prize for a paper on the lighting of the streets of a large city and Lavoisier decided to compete. He made an elaborate investigation, both practical and theoretical, of various methods of illumination; of the relative cost of lighting Paris with candle or oil lamps, of the advantage of various types of reflection, of the time of burning of different kinds of oil, and of the use of dissolved resin to prevent the freezing of oil in winter.

The prize was divided between three manufacturers, but Lavoisier's paper, an early and admirable piece of industrial research, was awarded a gold medal given by the King and was published by the Academy.[4]

ACADÉMIE DES SCIENCES

1768 was the fateful year of Lavoisier's life. In February the great promise he had shown as an investigator was recognized by his election as a member of the Academy of Sciences. From the day he joined that august body as 'adjoint chimiste' his wide knowledge and experience, his diligence, and his ability in drafting memoranda made him one of its most active members. Throughout his life a considerable portion of his time was devoted to its service, and Lavoisier was a most staunch defender of its privileges. He became a full member in 1778, director in 1785, and treasurer in 1791 until its suppression in 1793. The Government was constantly remitting questions to the Academy for reports, and Lavoisier's was the guiding hand in their preparation. He wrote over fifty memoranda on subjects ranging over The Great Frost of 1776, Prison Reform, Mesmerism, Water-Divining, the Creuzot Works, and The Development of Balloons.[5]

THE FERME

A few days after he had joined the Academy he became a member of the Ferme, the financial corporation that leased from the Government for periods of six years the privilege of collecting the indirect taxes. The annual payments had to be made in advance, so that membership of the Ferme involved a considerable investment of capital, the return on which depended on the ability of the administration and the state of trade. For Lavoisier it was not merely an investment, as he threw himself into the task of administration with his usual energy. His practical ability and his conscientious management soon made him one of the leading members of the Corporation. Lavoisier's scientific friends shook their heads at his new venture, but said cynically 'Well, at any rate he will give us better dinners'.

The collection of customs duties, octroi, and salt tax and the sale of tobacco, involving an annual income approximating

200 million livres, required an organization of nearly 30 000 officials working all over France. This was supervised by a number of committees drawn from the sixty members of the Corporation, their work involving a series of visits to the provinces. Lavoisier served on some of these, and he was specially charged with inspection of the tobacco factories and of the frontier guards to prevent smuggling. These journeys occupied a large part of his time from 1769 to 1771.

MARRIAGE

In 1771 Lavoisier married the daughter of Jacques Paulze, one of his colleagues in the Ferme. His wife was only fourteen years old, but she was a girl of character and intelligence and devoted her life to helping her husband (see Plate 2). She learnt languages and translated Priestley and Cavendish for him, and later she published a translation of the essays of Kirwan. She was constantly in the laboratory, and many pages of the records of his experiments are in her handwriting. She studied painting with David, and two drawings showing Lavoisier's experiments on respiration with Seguin are by her hand (see Plate 3). She also drew and engraved all the plates for his *Traité Élémentaire de Chimie*. Benjamin Franklin, in writing to thank her for his portrait, said: 'It is allowed by those who have seen it to have great merit as a picture in every respect; but what particularly endears it to me is the hand that drew it.'

We get a glimpse of her in Arthur Young's Diary in 1787.[6] 'To Monsieur Lavoisier by appointment. Madame Lavoisier, a lively, sensible, scientific lady, had prepared a *dejeuné Anglois* of tea and coffee, but her conversation on Mr Kirwan's *Essay on Phlogiston*, which she is translating from the English, and on other subjects, which a woman of understanding, that works with her husband in his laboratory knows how to adorn, was the best repast.'

Gouverneur Morris, the United States Minister in Paris, had to discuss with Lavoisier the price the Ferme should pay for Virginian tobacco and also a debt that was owing by his Government. His *Diary of the French Revolution*[7] gives us a picture of Lavoisier as a business man and of Madame Lavoisier as his hostess:

2 June 1789: Return Home, and then call on Monsr. Lavoisier . . . Talk of a future Contract. Thinks it should exist for a Part of the Supply. I agree with him and add that I would rather contract for the best Quality and leave the inferior Kinds to the Chance of Marketts. He asks at what Price. I tell him that what I say must not draw to any Consequence but be considered merely as Conversation. That it may be from 32₶ to 33₶. He thinks this much too high.

8 June 1789: Dine with Mr. de Lavoisier; as I am leaving he tells me that the Farm are determined to stand Suit and that he is sorry for it. Madame appears to be an agreable Woman. She is tolerably handsome, but from her Manner it would seem that she thinks her forte is the Understanding rather than the Person.

26 June 1789: Return Home and take a little Medicine and then go to Dinner at Mr. Lavoisier's. They are just returned from Versailles and Madame gives me the News. . . . At Dinner, Mr. Lavoisier, who seems to be a sensible and well informed Man, tells me that the usual Produce of France is from 5 to 6 Times the Seed sown.

25 September 1789: Go to the Opera according to my Promise and arrive towards the Close of the Piece at the Loge of Madame Lavoisier. . . . Go to the Arsenal and take Tea with Madame Lavoisier *en attendant le Retour de Monsieur* who is at the Hôtel de Ville. As Madame tells me that she has no Children I insist that she is *une Parasseuse*, but she declares it is only Misfortune. . . . Monsieur comes in and tells us of the *Obstination* of the Bakers. This Corps threatens the Municipality of Paris with a Discontinuance of their Occupation unless a Confrère, justly confined, is released. Thus the new Authority is already trampled on. I take Mr. Lavoisier apart and propose to him to negotiate with the Farm for the Debt of the United States. I offer him also a Concern. He will I think accept it, tho he rather objects. He wishes to know my Terms. I speak vaguely.

25 July 1791: Call on Mr. Franklin and we go together to Made. de Lavoisier's to Dinner. As the Party is rather High-Flying they conclude Haphazard that the Riot at Birmingham has been occasioned by the Government. This is ridiculous enough.

7 November 1791: Mr. Franklin dines here, and we go together after Dinner to see Made. Lavoisier where there are a Number of *Gens d'Esprit* who are in general but so so Company.

Few men can have been more happily married than Lavoisier. His wife shared in all his activities, she helped him to fight his scientific battles, and in the dark days to come she risked everything to try to save him.

EARLY CHEMICAL RESEARCHES

Even before his election to the Academy Lavoisier's interests were turning from geology to chemistry. His first two chemical papers were on gypsum, its composition, and its solubility in water.[8] Lavoisier used a hydrometer to show that the solution of gypsum was denser than water, and this led him to a systematic study of the subject of hydrometry in which he devised a new form of constant-immersion hydrometer, a small portable version of which he used to examine the densities of a large number of natural waters during his geological tour, in order to determine their content of dissolved salts. His practical mind saw the importance of water supplies to cities, and he was a strong supporter of de Parcieux's scheme for improving the water supply of Paris, on which he read papers to the Academy in 1769.

In order to standardize his hydrometers Lavoisier studied the effect of repeated distillation on its density and also the variation of density with temperature, in which he observed the increase of density to a maximum above the freezing-point. He was unaware that this had been discovered in Italy in the seventeenth century.

In 1768 he began to investigate the supposed conversion of water into earth which had been a matter of so much controversy since the time of Boyle. Lavoisier's method of attack is characteristic of his experimental methods.[9] He weighed a retort, introduced into it a known amount of fresh rain water, and luted on a closed receiver to act as a reflux condenser, after heating the retort to expel some of the air. The whole apparatus was then carefully weighed and the retort heated for 110 days in an oil-bath to about 80°C. A deposit of earth slowly formed in the retort, but the weight of the apparatus at the end was unchanged, showing that it had gained no fire material nor lost any water through the glass. This must have had an important influence on Lavoisier's ideas by strengthening his tacit assumption of the conservation of mass, and by showing him that in chemical changes involving exposure to heat the 'fire material' did not necessarily, as was often supposed, cause any change of weight. The solid was then collected, the water evaporated and the weight of the total dry solid residue was compared with that of the loss of weight of the empty retort during the operation. The solid weighed 20·4 gr, while the retort had lost only 17·4 gr. Lavoisier could not explain the greater weight of the solid residue (caused, as Meldrum has pointed out, by the absorption by alkali of carbon dioxide from the air), but he concluded rightly that in spite of this discrepancy the solid must have come from the glass and not from the water.

This was the most significant of Lavoisier's early papers, and it was followed early in 1772 by an investigation with Macquer and Cadet of the effect of heat on diamonds.[10] The varying resistance of gem stones to heat had excited the curiosity of experimenters for centuries, and the causes of the loss of weight of the diamond on heating had been continuously a matter for inquiry. Was it due to true volatilization as Boyle thought, or to a species of combustion as Macquer's recent observations indicated, or to a decreptitation into small fragments caused by contact with cold air? Those were the rival views which Lavoisier and his friends set out to investigate by comparing the effects of heating diamonds in open and closed vessels. Heated in an open retort to a high temperature for some hours they were found to lose weight and become discoloured. On the other hand, diamonds heated in a clay pipe filled with charcoal and closed with a lute, the pipe itself being enclosed in a nest of three crucibles with a similar lute to exclude air, lost no weight and showed only a slight superficial darkening and no loss of polish after exposure to the highest temperature available. Hence they concluded that the loss of weight depends on contact with the air and is due either to combustion or to some mechanical effect of the air in causing decrepitation. These experiments were continued later by Lavoisier using the great burning glass of Tchirnhausen, but the paper was not printed until 1776, and it is not clear at what date it was presented to the Academy.[11] Lavoisier's notebook shows that these experiments were still in progress in 1773.

COMBUSTION

The experiments on diamonds and probably papers by Sage and Cigna on the burning of phosphorus gave Lavoisier an interest in combustion, and his knowledge of the work of Boyle

Antoine Laurent Lavoisier

and Hales shows that he was studying the literature. At this time Stahl's theory of phlogiston still held the field. It was generally accepted by chemists that substances are inflammable because they contain this mysterious principle, which was liberated with heat and light so long as air was there to remove it. Respiration was thought to be a similar process, the air removing phlogiston from the lungs. It was a looking-glass hypothesis, a relic of alchemy, but it had an astonishing hold on chemists' minds until Lavoisier broke it.

The critical moment for Lavoisier came on 10 September 1772[12] when, having bought 1 oz of phosphorus, he found it could be burnt in glass vessels without breaking them and began to examine the absorption of air already noticed by Cigna and others during its combustion. On 20 October he sent a sealed note to the Secretary of the Academy giving the results of his experiments which showed that air is absorbed when phosphorus is burnt and that the phosphoric acid weighs more than the phosphorus.[13] On 1 November he sent another note describing his discovery that sulphur, like phosphorus, gains in weight during combustion, and that this increase of weight is due to a prodigious quantity of air which is fixed during combustion.[14]

Cette découverte, que j'ai constatée par des expériences que je regarde comme décisives, m'a fait penser que ce qui s'observait dans la combustion du soufre et du phosphore pouvait bien avoir lieu à l'égard de tous les corps qui acquièrent du poids par la combustion et la calcination; ... L'expérience a complétement confirmé mes conjectures; j'ai fait la réduction de la *litharge* dans les vaisseaux fermés, avec l'appareil de Hales, et j'ai observé qu'il se dégageait, au moment du passage de la *chaux* en métal, une quantité considérable d'air, et que cet air formait un volume mille fois plus grand que la quantité de *litharge* employée. Cette découverte me paraissant une des plus intéressantes de celles qui aient été faites depuis Stahl, j'ai cru devoir m'en assurer la propriété, en faisant le présent dépôt entre les mains du secrétaire de l'Académie, pour demeurer secret jusqu'au moment où je publierai mes expériences.

It was a moment of great elation for Lavoisier. To his practical mind increase in weight meant gaining something, not losing it, as the phlogistics argued. He saw a vision of something much bigger than a theory of combustion, a revolution in chemistry and physics. He saw, too, that its realization depended on a detailed investigation of the gases that were given off or absorbed in the chemical changes. It was no easy task, as Lavoisier had had comparatively little chemical experience, but he attacked it with courage on a wide front, and the next twelve months saw the most concentrated and sustained scientific work of his career. He was confident of his objective, and that confidence kept him going through the perplexities and disappointments of twelve crowded years.

OPUSCULES PHYSIQUES ET CHIMIQUES

The record of his next year's work was published in a separate volume, *Opuscules physiques et chimiques*, as it was too long for for *Memoirs* of the Academy.[15] It is described in the preface as the first of a series of investigations covering the wide field that he already had in view. It was Lavoisier's apprenticeship to the study of gases. He began with a careful study of the literature, and half the volume consists of a history of previous work. Hales, Black, and Priestley are the three chemists to whom he pays special attention. Through Hales he learned the experimental methods of John Mayow, and his own apparatus, though more elaborate, is obviously based on that of Hales. His systematic mind appears in the tables in which he assembles all the quantitative results of Hales's ingenious but almost random experiments. From Black he learned perhaps more than he said. Black's paper on the alkalis was a quantitative study after his own heart, and broadly speaking he was to do for metals and oxides just what Black had done for mild and caustic alkalis.

Having finished his survey of the literature, on 22 February 1773 he began his laboratory notebook with a remarkable forecast of the work it was to record:[16]

Qui m'a paru fait pour occasionner une revolution en physique et en chimie. J'ai cru ne devoir ne regarder tout ce qui a été fait avant moi que comme des indications: je me suis proposé de tout répéter avec de nouvelles précautions, afin de lier ce que nous connaissons sur l'air qui se fixe, ou qui se dégage des corps, avec les autres connaissances acquises et de former une théorie.

Lavoisier's own experiments, which are described in the second half of the volume, consist mainly of careful measurements of the changes in weight and volume that accompany the evolution or absorption of gas in the formation or decomposition of carbonates, the neutralization of acids, the reduction of metallic oxides, and the calcination of metals. His observations were thus concerned only with carbon dioxide and oxygen, but at that time Lavoisier had formed no definite view as to the nature of different gases. He was convinced of their fundamental importance and his aim was to bring them all into a common theory.

The experimental part of the paper begins with a repetition and testing of Black's work, using the results to establish the composition of quicklime, slaked lime, and chalk. This was then extended to the carbonates and hydroxides of the alkali metals and ammonium, and to the precipitation of metallic solutions by alkalis. The amounts of gas evolved by heating metallic calces with charcoal were measured and the gas was shown to be identical with that from chalk, but owing to faulty experimental methods the gas in each case was mixed with air, which must have complicated Lavoisier's conclusions. Finally, he measured the absorption of air when metals are calcined or phosphorus burnt in bell-jars over water.

Lavoisier's most important conclusion was that in confined spaces only a limited amount of calcination takes place, the increase in weight being roughly proportional to the volume of air absorbed. Thus it followed that metals could only fix part of the air, and the same result was obtained with phosphorus. But no conclusion is hazarded as to the nature of the gas which supports combustion, and the last sentence of the paper reveals one of the difficulties that must have been puzzling Lavoisier's mind. He says:

J'ai été curieux, relativement à des vues dont je rendrai compte dans un autre temps, d'observer si le mélange d'un tiers de fluide élastique des effervescences corrigerait l'air qui avait servi à la combustion du phosphore, et lui rendrait la propriété d'entretenir les corps enflammés. Le mélange fait, j'en ai rempli un bocal étroit et j'y ai introduit une bougie, mais elle s'y est éteinte sur-le-champ.

The paper as a whole is such a clear exposition of a logical sequence of quantitative experiments that it is disappointing that it should end on so inconclusive a note. Luckily we have in

Lavoisier's notebook the clue to his perplexities. Misled perhaps by Black's name of 'fixed air' for carbon dioxide (which he abandoned in the following year), Lavoisier thought that the gas which was 'fixed' in alkalis was identical with the gas 'fixed' by metals in their combustion to form calces. This is proved by several entries in his notebook; for example, on 1 July 1773 he wrote: 'Persuadé que la combustion du phosphore absorbe l'air fixe contenu dans l'air, ou plutôt le soupçonnant, j'ay pensé qu'en rendant de l'air fixe à cet air, on pourrait peut-être le rendre air commun.' He tries, of course, without success, but fails to push the matter further.[17]

A comparison of the notebook with the paper is interesting in another way. The experiments seem to have been made without any planned sequence, but in the paper the results are marshalled in perfect order, showing the systematic grasp Lavoisier already had of the whole field, his doubts only being recorded in the last sentences. It was a remarkable pioneer achievement in experimental technique, but progress had to wait until Lavoisier knew the relationship between carbon dioxide and oxygen. He had still much to learn, despite his brilliant intuition.

CALCINATION OF METALS

Lavoisier was unshaken in his belief that increase in weight meant chemical combination, and immediately the *Opuscules* was published he began experiments on the calcination of tin and lead in sealed vessels. Boyle had done similar experiments but had only reweighed the vessels after opening them, when he found an increase in weight which he explained by the passage of fire material through the glass.[18] Lavoisier saw the opportunity for a crucial experiment by weighing the vessels both before and after they were opened. A summary of his results was deposited with the Academy on 24 April 1774, but was not read until the public session on 14 November,[19] following Lavoisier's usual plan of keeping a tit-bit for that occasion. Without giving any numerical details Lavoisier said that in sealed vessels the calcination of lead and tin ceased after an hour's heating, the amount of calx formed being greater in larger vessels. In every case the weight of the vessel was unaltered if it was weighed before it was opened. When the seal was broken, air was heard to rush in, and the weight of the vessel was then greater, the increase being 'exactly proportional to its capacity'. Hence it was clear that the increase in weight was due not to the passage of fire-stuff through the glass but to something the metal had 'borrowed' from the air in the vessel which had converted it into a calx.

Lavoisier ended with a significant sentence about the nature of the air that remained after metal had been calcined in it:

Cet air ainsi dépouillé de sa partie fixable (je pourrois presque dire de la partie acide qu'il contenoit); cet air, dis-je, est en quelque façon décomposé; et il m'a paru résulter de cette expérience un moyen d'analyser le fluide qui constitue notre atmosphère; et d'examiner les principes qui le constituent. Quoique je ne sois pas arrivé à cet égard à des résultats entièrement satisfaisans, je crois cependant être en état d'assurer que l'air aussi pur qu'on puisse le supposer, dépouillé de toute humidité et de toute substance étrangère à son existence et à sa composition, loin d'être un être simple, un élément, comme on le pense communément, doit être rangé au contraire tout au moins dans la classe des mixtes, et peut-être même dans celle des composés.

Lavoisier found that his experiments were not as new as he had thought, and he added a note saying that Beccaria had made the same observations on tin fifteen years earlier, and that Priestley had pointed out the limited extent of calcination in sealed vessels. But the ultimate value of an observation lies in the deduction from it, and here Lavoisier alone saw the truth.

A full account of these experiments was not given until 1777, when Lavoisier had to admit that owing to their difficulty only two with tin were completely successful and none with lead.[20] However, the numerical results of the two successful experiments showed the care and skill with which they had been carried out, and justified the conclusion that the weight of the sealed vessels was unaltered, within the errors of experiment, after calcination had occurred. By then Lavoisier had further evidence about the nature of air, which he describes as a mixture of a salubrious portion which combines with metals during calcination, leaving 'une espèce de mofette' which will not support life or combustion, and is itself 'fort composé'. From his results he suspected that the former is the denser of the two, air being intermediate. His experiments were, in fact, sufficiently accurate to justify this conclusion.

These experiments on the calcination of tin must have strengthened Lavoisier's confidence in the correctness of his theory, and from March 1774 onwards his notebook shows that he was widening his front of attack, for combustion was only one aspect of the revolution he had in mind.[21] This year saw the beginnings of work on a number of problems whose solution eluded him for the moment, as the background of his knowledge was as yet insufficient. In each there was a quantitative approach to determine both the composition of substances and the nature of the chemical changes that they undergo.

He begins by burning inflammable air (hydrogen) which ought on his theory to increase in weight, and he was puzzled to find a loss owing to the escape of water vapour. This was a problem that continued to perplex him for nine years.

Next came a study of the formation of nitrous ether, also abortive, then an attempt to determine the composition of nitre by exploding a mixture of it with charcoal, and analysing the gaseous products. Knowing the percentage of caustic alkali in the nitre he tried to calculate its composition, but he did not know how much the charcoal contributed to the weight of fixed air.

In October he is repeating a number of Priestley's experiments on various gases, including his production of an acid by passing electric sparks in air. But it was characteristic of Lavoisier's lack of qualitative intuition that in spite of his interest in acids he does not follow up this clue, and the explanation had to wait for Cavendish ten years later.

Next come some revealing thoughts on 'vegetable analysis' and plans for future work, the birth of organic analysis.

Nous ignorons: 1° Quelle est la qualité de cette immense quantité d'air qui se dégage pendant la distillation: il y a probablement de l'air fixe et de l'air inflammable; 2° ce que c'est que l'huille: il paraît que par la combustion on peut la réduire en air et en eau; mais nous ne savons rien au de-là. Déterminer les proportions des deux en brûlant une lampe en un vase clos. 3° Ce que c'est que le charbon. Nous savons bien qu'en brûlant il convertit l'air évidemment en air fixe: mais nous ne savons pas s'il donne lui-même de l'air fixe.

Lavoisier is still not quite free from the incubus of phlogiston, and he suggests burning charcoal in a sealed vessel to see if any phlogiston escapes through the glass during combustion with a corresponding loss of weight.

PRIESTLEY AND OXYGEN

Lavoisier's notebooks show that in 1774 his mind was roving over a wide range of problems of great significance but too complex to be solved with his present knowledge. He was still puzzled about the relation of fixed air to the air absorbed in calcination, and had tried to find a calx which could be reduced and give up this air in the absence of charcoal. It seems clear that Priestley gave him the clue he needed when he was in Paris in 1774. On 1 October he told Lavoisier that by heating the calx of mercury he had got a gas in which a candle burnt particularly brightly. It was not until the following 1 March that Priestley realized the remarkable properties of this gas and thus discovered oxygen. Lavoisier got some calx of mercury from Cadet, and after some preliminary experiments in November three months elapsed until he completed them in three days' work with Trudaine at Montigny. He described the results at the public session of the Academy in April 1775 in a paper on 'The nature of the principle that combines with metals during their calcination and augments their weight'.[22]

His main interest was to compare the fixed air he got by heating the calx of mercury with charcoal with the gas he got by heating it alone, which to his surprise was quite different and 'more respirable, more combustible and consequently purer than ordinary air'. His difficulty about the relation of these two gases was therefore solved. But qualitative experiments were never Lavoisier's strong suit, and in this paper for once he is inconsistent. Perhaps it was written hurriedly for the public session. At one point he describes the 'principle' absorbed in calcination as 'neither one of the constituents of the air, nor a particular acid existing in it, but the air itself without alteration or decomposition'. And he has the curious idea that the air can be purified whenever it is absorbed in a calx and then set free again. In another part of the paper he says that the 'principle' is the purest part of the air. Lavoisier is rarely guilty of such inconsistency, but in this paper he clearly failed to realize the full significance of the facts he had observed.

The paper was published in Rozier's *Journal* and soon drew a criticism from Priestley,[23] whose observations were much more accurate:

> Having mentioned the paper of Mr Lavoisier's ... I would observe, that it appears by it, that, after I left Paris, where I procured the *mercurius calcinatus* abovementioned, and had spoken of the experiments that I had made, and that I intended to make with it, he began his experiments upon the same substance, and presently found what I have called *dephlogisticated air*, but without investigating the nature of it, and indeed, without being fully apprised of the degree of its purity. ... He therefore inferred, as I have said that I myself had once done, that this substance had, during the process of calcination, imbibed atmospherical air, not in part, but in whole. ... As a concurrence of unforeseen and undesigned circumstances has favvoured me in this inquiry, a like happy concurrence may favour Mr Lavoisier in another; and as, in this case, truth has been the means of leading him into error, error may, in its turn, lead him into truth.

It is curious that Lavoisier failed to see at once that the experiments with the calx of mercury gave the support he needed for his theory, and that he was so slow to complete them. From May until September 1775 his laboratory notebook is a blank. The reason for all this was a new preoccupation in his mind.

RÉGIE DES POUDRES

Turgot, a friend of Lavoisier, had just been made Controller-General, the one man who might perhaps have saved France from a Revolution, if vested interests had not been too strong. Lavoisier was in touch with every side of the administration, and his critical eye had noted the inefficiency of the national production of gunpowder which was made under contract by a financial company. Lavoisier saw that their methods of collecting nitre and of making gunpowder were costly and wasteful, and that production was insufficient. He remembered, too, that the prohibitive cost to France of imported powder during the Seven Years' War was one cause that had led to the peace of 1763. He succeeded in persuading Turgot that it would not only pay the government to cancel the existing contracts with compensation and take over the manufacture itself, but that it was essential for the national safety. This was done by a decree in March 1775, and Lavoisier and three of the former officials were put in charge of the powder factories.

Lavoisier at once concentrated all his energies on the new task. He moved to the Arsenal, where he lived and had his laboratory until 1792. He then began a thorough investigation into the methods of making and collecting nitre, nitrification being the only source of saltpetre apart from some deposits which he spent three months in surveying. His laboratory notebooks for 1775 and 1776 are full of records of various investigations into the manufacture of nitre and gunpowder.

In 1777 he drew up comprehensive instructions for the construction and operation of nitrification plants embodying many improvements, which represented a fine piece of chemical engineering.[24] With his usual eye to finance he calculated that the plants should yield a net revenue of 15 per cent on the capital. Great care was taken to select suitable staff, who had to pass an examination in chemistry, mathematics, and the construction of powder mills.

The results were even better than Lavoisier had predicted. In three years the improvement of the powder had increased the range of French weapons by 60 per cent. Production increased steadily, and by 1788 there was a reserve of 5 000 000 lb. in the magazines, and the economies had amounted to 20 000 000 livres.

However, the effect of Lavoisier's work did not end with the Napoleonic wars. He had interested young Irénée Du Pont, the son of the physiocrat, in chemistry, and later he became Lavoisier's assistant at the Arsenal. Irénée and his father were constitutionalists, and in 1792 they were fighting beside the Swiss Guard at the Tuileries. Saved by a miracle, they emigrated later to America. There on the banks of the Brandywine River at Wilmington, Irénée Du Pont built another powder factory destined twice to become a great arsenal of the Allied Nations.

THE DISCOVERIES OF 1776 AND 1777

In September 1775 experiments in the laboratory were resumed, though most of the work in the autumn was concerned with the manufacture of nitre. But gradually Lavoisier began to take up his scientific work again. He dissolves mercury in nitric acid, evaporating the solution to dryness and measuring the gases evolved when the mercuric nitrate is heated. This he uses as a method of analysing nitric acid, as the experiment ends with the same weight of mercury as at the start. From the proportions of nitrous gas and pure air he calculates the percentage composition of nitric acid, assuming the other component to be water. The results were read to the Academy in 1776 in a paper 'On the existence of air in nitric acid',[25] which he takes as an illustration of his general thesis that 'non-seulement l'air, mais encore la portion la plus pure de l'air, entre dans la composition de tous les acides sans exception; que c'est cette substance qui constitue leur acidité'. He shows how he gets a mixture of nitrous air and pure air from nitric acid, and how by recombining them he gets nitric acid back again. He always tried to verify his conclusions both by analysis and synthesis.

In this paper he twice pays a tribute to Priestley as the original discoverer of many of the facts he describes: 'comme les mêmes faits nous ont conduits à des conséquences diamétralement opposées, j'espère que, si on me reproche d'avoir emprunté des preuves des ouvrages de ce célèbre physicien, on ne me contestera pas au moins la propriété des conséquences.'

This was the only paper he published in 1776, but it was a very busy year in the laboratory. Many of Priestley's experiments was repeated. Work were in progress on the analysis of air by the calcination of mercury, on the composition of various acids, on respiration, on the burning of a candle, on the latent heat of vaporization, and on the heat produced in the neutralization of acids.

Lavoisier is now quite clear about the nature of the air and of combustion, and for the first time he openly challenges the theory of phlogiston. During 1777 the evidence he had accumulated in support of his theory was presented to the Academy in nine papers, one of which was not read until 1779, and none was printed until 1780. Air, he says, is a mixture, four-fifths

Antoine Laurent Lavoisier

being an inert gas 'mofette' which takes no part in respiration or combustion, and one-fifth a gas 'eminently respirable', a constituent of all acids, which he therefore named oxygen. The compositions of carbonic acid, nitric acid, phosphoric acid, sulphuric and sulphurous acids had been determined. Respiration and the burning of a candle both convert oxygen into carbonic acid or 'acide crayeux aeriforme', the name Lavoisier has substituted for fixed air. Slow combustion in the lungs is suggested as the source of animal heat. Finally, the existence of the three states of matter is shown to depend on their heat content, and the heat of combustion is explained by the heat given out by gaseous oxygen in its change of state and combination.

Supported by this evidence Lavoisier attacks the theory of phlogiston, and shows that his own explanation is much simpler, involving no unwarranted assumptions such as the presence of an immense quantity of fire material in solids like the diamond, or that substances gain in weight while losing part of their contents.

It was a year of great achievement; the problem of combustion was solved, but Lavoisier's theory gained no adherents. It was incomplete, and there were many other simple reactions he could not explain. He was still puzzled by inflammable air, and he had not tried to discover the nature of the inert constituent of the atmosphere.

LAVOISIER'S EXPERIMENTAL FARM

The clue to both problems was to come eventually from Cavendish, and Lavoisier's notebooks of 1778–82 contain no records of new and important discoveries. His hands must have been full enough with the Arsenal, the collection of taxes and the business of the Academy. To these he now added a fresh activity—an experimental farm. On his many journeys through France Lavoisier, like Arthur Young, had seen the poor state of French agriculture and the hard lot of the peasants. The staple crop was wheat grown three years out of four, and the head of livestock was too small to yield the manure required for good cultivation. Lavoisier saw the advantages of the British system of mixed farming, with its larger capital investment, more livestock, and wheat grown only one year in four.

DHC

So in 1778 he started an experimental farm at Fréchines near Blois, in a district where the standard of farming was low, in order to investigate and demonstrate the possibilities of improvement. Lavoisier always took the big view, and his aim now was to increase the wealth of France and to ease the life of the peasants.

The experiments were carried out as far as possible with the same strict control as if they had been made in the laboratory. He kept records at Paris of every field, its size, soil, crop, and yield, with an annual debit and credit account.

His main objective was to increase the amount of fodder crops and raise the number and quality of his livestock. He had some disappointments at the start—lucerne failed and clover did badly in dry years, but sainfoin was a success. He introduced ley farming on the fallows, catch crops and root crops, and gradually increased his livestock. The folding of sheep was another successful innovation. No point was too small for Lavoisier, and one of his papers describes in great detail a simple way of making hurdles.

After nine years' development the farm was in much better shape, but Lavoisier was disappointed at the slowness of progress, and he saw no prospect of getting a 5 per cent return on his investment.[26] This and the shortness of the leases explained why French farming suffered so badly from a lack of capital. However, that was not the end, as by 1793 his wheat crop had doubled, his stock had multiplied five times, and his neighbours were copying his improved methods of farming.

When Calonne appointed a Consultative Committee on Agriculture in 1785, Lavoisier was its secretary and drafted its reports. Among its schemes were projects for experimental farms, a museum of agricultural implements, and a school of textiles made from home-grown materials, which was actually started in Paris. Among Lavoisier's friends were the leading physiocrats, including Du Pont de Nemours and Malesherbes. While he sympathized with their views on freedom of trade, and regarded agriculture as the primary source of national wealth, he recognized that industry, too, had its place in the national economy. One of his memoranda dealt with the bad effects of restrictive regulations on agriculture. Another, dealing with the many relics of feudalism which handicapped the French

Antoine Laurent Lavoisier

farmer, reads like pages from Arthur Young's Diary. A third, 'Instructions to the Provincial Assemblies for improving agriculture', recommends many of the new practices Lavoisier had tried himself at Fréchines. In the end, however, little was done owing to the lack of interest of the ministers.

LAVOISIER'S DAY

Lavoisier's immense output of work in so many fields during this period was only made possible by his quickness of mind and memory, his power of concentration on the matter in hand, and his methodical habits. While he was living at the Arsenal 6 hours a day were given to science, in the early morning from 6 o'clock to 9 and from 7 o'clock to 10 after dinner. The rest of the day was spent in dealing with the business of the Ferme or the powder factories, in the meetings of the Academy or of the many Commissions of which Lavoisier was a member. Sunday was the happiest day of the week, as he spent the whole of it in his laboratory, which soon became a rendezvous for intellectual society in Paris. Scientists, ministers, economists, and distinguished visitors from abroad went there to see Lavoisier's latest experiment and to join in the discussion of its significance. This was the usual preliminary to one of his papers at the Academy. The younger men interested in science were equally welcome with their seniors, and they often helped Lavoisier with his experiments. The memories of those brilliant gatherings in his laboratory, which played no unimportant part in scientific history, remained fresh in the minds of those who shared in them.

THERMOCHEMISTRY

One of Lavoisier's great services to chemistry was to disentangle the chemical and physical aspects of chemical change. The phlogiston theory had been responsible for much loose thinking, since it assumed that this mythical principle could modify both the chemical and the physical properties of matter. Lavoisier concentrated first on the alterations in weight accompanying chemical change, particularly combustion. His belief in the conservation of matter was strengthened by his experiments in sealed vessels which showed no evidence of changes in weight due to the gain or loss of fire material. This was confirmed by unpublished experiments which proved that the weight of sealed tubes of water remained unchanged when they were frozen. The distinction between chemical and physical changes simplified his task greatly, and by 1777 his evidence on the gravimetric side was sufficient to enable him to give a satisfactory explanation of the nature of air and its role in the chemical changes involved in combustion and respiration.

Lavoisier was now striving to give a similar picture of the physical changes accompanying chemical action. He recognized the dependence of the three states of matter on their heat content, and he explained the heat of combustion by the change in the heat content of the oxygen concerned. He had no idea of the conception of energy, and he did not connect the evolution of heat with the affinities of the elements for each other. The difference in the heat of neutralization of caustic and mild alkalis (hydroxides and carbonates) he explained by the absorption of heat in the liberation of carbon dioxide as a gas.

Lavoisier had always been interested in thermometry and heat changes since the start of his scientific work. In 1772, when Desmarest told the Academy of Black's work on latent heat, Lavoisier described an experiment made the preceding year in which he had discovered independently the same phenomenon, finding to his surprise that on mixing water and crushed ice the temperature of the water remained at $0°C.$, until all the ice was melted.[27] In 1773, when studying the crystallization of sodium sulphate by cooling solutions, he observed the temperature arrest which occurred when crystallization began and lasted until it had finished. This he explained correctly by the heat of crystallization of the salt.[28]

In his physical work he had the advantage of partnership with Laplace, who worked in his laboratory at intervals from 1777 to 1785. In 1777 they were measuring the vapour pressure of liquids, and determining latent and specific heats by the method of mixtures. Black's unpublished work was known to them, and de Luc and Crawford were working in the same field. In 1782, finding the method of mixtures unsatisfactory for measuring heats of reactions, they invented a new method, the ice calorimeter, in which the amount of heat given out is measured by the amount of ice melted (Plate 3). The calorimeter consisted

Antoine Laurent Lavoisier 41

of two concentric vessels each containing crushed ice, the purpose of the outer vessel being to insulate the inner one against any gain or loss of heat from outside. There was a space in the inner vessel to hold the substance or reaction under examination, the amount of heat enclosed being measured by the amount of water that ran down from the ice as it melted. It was a beautifully simple and accurate apparatus which they used to measure specific heats, including the specific heats of gases by a flow method, heats of reaction, heats of combustion of phosphorus, carbon, ether, and oil, and the heats of detonation of nitre mixed with charcoal or sulphur.

Thermochemistry dates from Lavoisier and Laplace's great paper 'Sur la Chaleur' of 1783,[29] and one of their generalizations is a partial anticipation of Le Chatelier's Theorem which came a century later: 'Dans les changements causés par la chaleur à l'état d'un système de corps, il y a toujours absorption de chaleur; en sorte que l'état qui succède immédiatement à un autre, par une addition suffisante de chaleur, absorbe cette chaleur sans que le degré de température du système augmente.'

More measurements were made in the following winter, including the heats of combustion of hydrogen and of wax, which Lavoisier showed was nearly equal to the sum of the heats of combustion of the hydrogen and carbon it contained.

The ice calorimeter thus gave Lavoisier precise data for the heat changes involved in the chemical reactions in which he had already determined the weight changes involved. He now had a complete picture of these two different aspects which enabled him to show how unnecessary it was to assume the existence of phlogiston, as everything could be explained much more simply without its aid.

In 1777 Lavoisier had explained the source of animal heat as being the heat evolved when carbonic acid (*air crayeux*) is produced in the lungs from the oxygen of the air during respiration. This was an entirely new point of view, as Haller had accepted Stahl's explanation that the heat of the body was due to the friction of the blood in the arteries, though he was baffled by the constancy of body temperature. The ice calorimeter gave the opportunity to verify this speculation by measuring the heat given out by a guinea-pig and comparing it with the heat evolved in the formation of the volume of carbonic acid which an animal of similar weight expired during the same period.

First the amount of carbonic acid expired by guinea-pigs in a period of 10 h was measured by absorbing it in potash bulbs. From the mean of their experiments Lavoisier and Laplace calculated that the combustion of the corresponding amount of carbon would have melted 10·38 oz of ice. A guinea-pig was then kept in a calorimeter for the same period and the heat given out by it melted 13 oz of ice. Part of the excess, they said, might have been due to the cooling of its extremities—it might well have had cold feet! From the results they concluded: 'la conservation de la chaleur (animale) est due, au moins en grande partie, à la chaleur que produit la combinaison de l'air pur respiré par les animaux avec la base de l'air fixe que le sang lui fournit.' The maintenance of a constant body temperature under different climatic conditions they ascribed to differences in the rate of evaporation of moisture rather than to differences in the amount of carbonic acid formed during respiration.

This was one of Lavoisier's most brilliant papers, remarkable for the beauty of the experimental technique, for the directness of attack, and for its verification of his theory of animal heat, which gave a new significance to metabolism. No doubt the paper owed much to Laplace, who was one of the first converts to Lavoisier's theories. The full significance of their eight years of partnership has not had the attention it deserves.

THE NATURE AND COMPOSITION OF WATER

His notebooks show repeatedly how puzzled Lavoisier still was as to the nature of the inflammable air he got in various ways, by the solution of metals in acids, or the distillation of vegetable substances. In 1774 he had tried and failed to ascertain the increase in weight when inflammable air was burnt. He expected inflammable air to give an acid on combustion but could find none. Since the inflammable air came from sulphuric acid and metals Lavoisier thought it should give sulphuric acid. Bucquet thought it should give fixed air. Experiments in 1774 and 1777 showed that neither acid was produced. He returned to the problem again in 1781 and 1782, when he and Gingembre burnt a jet of oxygen in inflammable air and found neither carbon dioxide nor any acidity, which

Antoine Laurent Lavoisier

still surprised Lavoisier 'que l'analogie m'avait porté invinciblement à conclure que la combustion de l'air inflammable devait également produire un acide'. Hence in 1782 he had no idea that water was the product.

Meanwhile other chemists were busy with the same problem, and Cavendish had found that when a mixture of inflammable air and common air is exploded in a closed vessel water is formed without any change in weight. When Blagden, the Secretary of the Royal Society, was in Paris, he told Lavoisier of Cavendish's experiments. On 23 June 1783, Lavoisier took two gas holders containing inflammable air and oxygen and joined them up to a jet at which the mixture could be burnt for a considerable time in a bell-jar over mercury. The first experiment was made by Lavoisier and Laplace in the presence of Blagden and a number of other scientists. A considerable quantity of pure water collected on the surface of the mercury, and although it was not possible to establish directly that the weight of the water found was equal to that of the two gases Lavoisier had no doubt of it. He wrote 'comme il n'est pas moins vrai en physique qu'en géométrie que le tout est égal à ses parties, de ce que nous n'avions obtenu que de l'eau pure dans cette expérience, sans aucun autre résidu, nous nous sommes crus droit d'en conclure que le poids de cette eau étoit égal à celui des deux airs qui avoient servi à la former.'[30]

The next day the experiment was described to the Academy. 'Nous ne balançâmes pas à en conclure que l'eau n'est point une substance simple, et qu'elle est composée poids pour poids d'air inflammable et d'air vital.' Lavoisier then heard that Monge had made similar experiments. Having determined the combining volumes of the two gases he had found that the weight of the water was almost equal to that of the two gases, the specific gravities of which he had determined.

Lavoisier must have seen in a flash the wide significance of this new view of the nature of water, which gave him the clue to so many of his unsolved problems. It explained the source of the water he got in organic combustions, and he saw the possibility of using them to determine the composition of organic bodies.

Additional evidence was obtained by studying the decomposition of water by long standing over cold iron, and by passing steam over hot iron or charcoal, as well as from experiments on its formation by passing inflammable air over heated metallic calces, thus extending some observations made by Priestley.

Lavoisier was naturally anxious to know the percentage composition of water, as it would enter into so many of his calculations. In his first paper to the Academy he gives the following figures based on experiments made with Meusnier to determine the combining volumes of the two gases and their densities:

	Livres
Air vital, ou plutôt principe oxygine	0·86866273
Air inflammable, ou plutôt principe inflammable de l'eau	0·13133727
Total	1·00000000

Lavoisier at this stage has a preference, like Boyle, for the word 'principle' rather than 'element'. The name of hydrogen was not adopted until the whole nomenclature was revised four years later.

There are numerous references to a new apparatus that was constructed to prove that the weight of water was equal to that of the gases from which it was formed, and to determine its composition. Arthur Young saw it in Lavoisier's laboratory in 1787. 'In the apparatus for aerial experiments', he writes in his Diary,[31] 'nothing makes so great a figure as the machine for burning inflammable and vital air, to make or deposit water; it is a splendid machine. Three vessels are held in suspension with indexes for marking the immediate variations of their weights; two that are as large as half hogsheads contain the one inflammable, the other the vital air, and a tube of communication passes to the third, where the two airs unite and burn. . . . If accurate (of which I confess I have little conception) it is a noble machine.'

It has been questioned whether Lavoisier ever used this apparatus successfully, as the results were never published by him. The experimental details, however, are given in the *Traité*. Hydrogen was burnt at a fine jet in a glass globe containing oxygen, both gases having been dried by passing over potash (Lavoisier says that calcium nitrate or chloride would have been better). The increase in weight of the globe together

Antoine Laurent Lavoisier

with measurements of the volumes of the gases consumed, and a knowledge of their densities, gave Lavoisier the data he needed. Details of one experiment were published in an unsigned article in the *Journal Polytype*,[32] an ephemeral scientific journal in which Lavoisier was interested. The weight of water found was within 1 per cent of that of the two gases, and its composition is given as 85 per cent oxygen and 15 per cent hydrogen which are the figures Lavoisier used subsequently in all his calculations. The hydrogen content was too high owing to errors in the determination of its density.

It was a very difficult experiment and a great test of Lavoisier's skill. The final solution had to wait until Morley's classical investigations just a century later.

Lavoisier lost no time in exploiting this new view of the composition of water. He saw that it explained the formation of water in organic combustions, and he was now able to use them to determine the composition of organic bodies. He determined the heat of combustion of hydrogen and saw that the oxidation of hydrogen as well as carbon in the body would account for more oxygen being consumed in respiration than the expired carbonic acid could account for. This would also explain why the amount of heat produced by an animal was greater than that calculated from its output of carbonic acid.

THE NEW CHEMISTRY AND THE REFORM OF NOMENCLATURE

The recognition of hydrogen as an element was a great advance. It explained many of Lavoisier's perplexities, and shortly afterwards Cavendish discovered that the inert part of the atmosphere, Lavoisier's 'mofette', gave nitric acid when sparked with oxygen, and Berthollet found that ammonia is a compound of 'mofette' with hydrogen.

The way was now clear. With the recognition of the elements oxygen, hydrogen, nitrogen (or azote as the French chemists called it), carbon, sulphur, phosphorus, and the metals, Lavoisier could explain quite simply the composition of chemical compounds, their reactions, and their quantitative relations. There were no longer obscurities to serve as a stronghold for phlogiston. Laplace and some of the younger physicists were already on his side. Of the chemists, Berthollet announced his conversion at the Academy in August 1785, followed soon by Fourcroy and Guyton de Morveau. The latter had published proposals for a reform of chemical nomenclature in 1782, but these were based on phlogiston. Discussions between the four chemists in 1786 led to their publication jointly of a new method of chemical nomenclature based on Lavoisier's theory[33] In the introduction Lavoisier acknowledges his debt to Condillac, the disciple of John Locke. From Condillac's *Logic* he had learnt the value of clarity of expression, of language as an instrument of analysis, of words as the symbols in the algebra of thought.

Till then phlogiston had been the only common basis. Names had had little reference to the nature or relationships of the substances that bore them, and many were relics of alchemy. Lavoisier's work had made a system possible, in which the names of elements like hydrogen and oxygen told something about them, and the names of compounds like sulphates and sulphites indicated their composition. The system devised by the four chemists, with few exceptions, is the one we use to-day, and its adoption helped greatly towards the general acceptance of the new theory.

'TRAITÉ ÉLÉMENTAIRE'

The nomenclature having been agreed upon, Lavoisier was drawn irresistibly, as he tells us, to use it to give a logical account of his new theory. His *Traité élémentaire de chimie*[34] was finished early in 1789. Its publication marks the birth of modern chemistry, and it threw a new light on all the related sciences. It was no ordinary textbook, as the first person singular appears on every page. It was in fact an autobiography of Lavoisier's work since 1772. The discoveries of seventeen years were compressed into a few pages, and the experimental basis of the new chemistry presented in such a clear compelling fashion that it won immediate acceptance. It was chemistry as Lavoisier saw it without any history or even a mention of phlogiston.

The first part outlines Lavoisier's theory of chemistry and gives the results of the crucial experiments on which it was based. 'C'est elle seule qui contient l'ensemble de la doctrine que j'ai adoptée; c'est à elle seule que j'ai cherché à donner la forme véritablement élémentaire.'

Antoine Laurent Lavoisier

After giving his views on the nature of heat, he describes the experiments which established the composition and chemical nature of the atmosphere, water, and a number of acids, gradually building up his picture on experimental results. He then sets out his views on organic chemistry, his experiments on fermentation, and his work on the analysis of organic compounds by combustion.

Oxygen is the central element, a constituent, Lavoisier thought, of all acids and all bases. Hence he predicted rightly the existence of the alkali metals isolated twenty years later by Davy. But his assumption of the presence of oxygen in all acids was less happy, and here again Davy established the facts.

The *Traité* was the forerunner of the textbooks of the nineteenth century, and some of Lavoisier's views, the dualistic nature of salts and the conception of compound radicals, were to have a profound influence on chemical thought and to be the issues on which many battles were to be fought.

The idea of the chemical equation had been implicit in all Lavoisier's quantitative work:

Car rien ne se crée, ni dans les opérations de l'art, ni dans celles de la nature, et l'on peut poser en principe que dans toute opération, il y a une égale quantité de matière avant et après l'opération; que la qualité et la quantité des principes est la même, et qu'il n'y a que des changemens, des modifications ... je puis considérer les matières mises à fermenter et le résultat obtenu après la fermentation, comme une équation algébrique.

The first time that Lavoisier actually used the familiar form of the chemical equation was in the chapter on alcoholic fermentation when he wrote:

Le moût de raisin = acide carbonique + alkool.

The second part of the *Traité* contains lists of salts with general methods for their preparation from each acid. It is a useful summary containing nothing 'qui me soit propre'.

The first part gave few experimental details: 'J'ai reconnu ... que des descriptions minutieuses ... figuroient mal dans un ouvrage de raisonnement; qu'elles interrompoient la marche des idées, et qu'elles rendoient la lecture de l'ouvrage fastidieuse et difficile.' In the third part Lavoisier gives a full account of the construction of the apparatus and the experimental methods which he had gradually developed. It is of intense interest in revealing the care and skill he devoted to his experiments, most of which were made with his own hands. Its value is enhanced by many illustrations drawn and engraved with loving care by Madame Lavoisier who had so often seen the apparatus in use in the laboratory. The book has the same individual quality as Faraday's *Chemical Manipulation*. It begins with an appeal for the use of the decimal system and ends with the prototype of future tables of physico-chemical constants.

The publication of the *Traité* marks the end of the phlogistic period and the beginning of modern chemistry. The clear logic of its presentation won immediately almost general acceptance. It was translated into several languages and became the accepted method of teaching. It had been a long fight, but in the end Lavoisier won an almost bloodless victory. In 1791 he wrote to Chaptal: 'Toute la jeunesse adopte la nouvelle théorie et j'en conclus que la révolution est faite en chimie.' Even Kirwan in 1792 wrote to Berthollet: 'Enfin, je mets bas les armes et j'abandonne la phlogistique.'

RESPIRATION AND METABOLISM

After the publication of the *Traité* Lavoisier's interests turned to organic and physiological chemistry. Almost his last scientific work was on respiration and metabolism. In 1785, in a brilliant lecture to the Society of Medicine, he described experiments on birds and guinea-pigs, showing that they could live for long periods in pure oxygen and that, to his astonishment, the rate of production of carbon dioxide was practically unaltered.[35] He found later that they could live equally well in an atmosphere in which hydrogen was substituted for the nitrogen of the air. He saw that water was formed as well as carbon dioxide in the process of metabolism, thus accounting for what is now called the 'respiratory quotient', and he was able to calculate the amount of water produced. The heat evolved in its formation explained why his measurements with Laplace had shown that an animal gave out more heat than could be accounted for by the heat of formation of the carbon dioxide it expired.

From his observations on the distress of animals breathing in confined spaces as the concentration of oxygen diminished and that of carbon dioxide increased, Lavoisier's practical mind travelled to the state of the atmosphere in crowded rooms. He measured its composition in hospital wards and in the Comédie Française. Finding a deficit of oxygen and an increase of 2 or 3 per cent of carbon dioxide, he urged the need for better ventilation. He calculated the rate at which the atmosphere deteriorated, producing that 'impatience machinale' in an audience which was such a sad handicap to the last speaker at a meeting of the Academy.

Finally, he saw the danger that emanations from the lungs might spread disease in crowded places, and he urged the need for knowledge of the ways in which infection spreads in order to protect the health of people in large towns. The lecture was a brilliant forecast of the problems of ventilation.

Experiments on a human being were the culmination of Lavoisier's work on respiration and the subject of his last two communications at public sessions of the Academy in 1789 and 1790.[36] Seguin, his co-author, was the experimental subject. Measurements were made of his consumption of oxygen, his output of carbon dioxide, his rate of respiration, and his pulse rate under different conditions. We know from sketches made by Madame Lavoisier the general experimental arrangement (Plate 3). The results revealed the main factors regulating metabolism: the temperature of the environment, the output of work, and the digestion of food. Lowering the room temperature from 80 to 61 °F raised the absorption of oxygen by a man at rest by 11 per cent/h, the digestion of food raised it by 50 per cent, doing work at the rate of 37 000 ft lb/h raised it by 160 per cent, and the digestion of food simultaneously with work at the rate of 39 000 ft lb/h raised it by 280 per cent. Lavoisier showed also that the increase in the pulse rate was proportional to the amount of work done, and that the amount of oxygen absorbed was proportional to the product of the pulse rate and the number of inhalations.

He saw at once the secret of the constancy of the body temperature under varying conditions of climate and occupation that had so long remained a mystery. It is regulated automatically by three balancing factors—respiration which produces heat by oxidation of carbon and hydrogen, transpiration which increases and diminishes as needed to remove any excess of heat by evaporation of water from the skin, and digestion which replaces the matter lost in the first two processes. The directness of the attack, the swift recognition of the wide implications of the results, are characteristic of Lavoisier at his best.

Finally, he discusses their bearing on the state of the body in health or disease. Fevers he thought were nature's method of restoring any disturbance of the equilibrium. The art of the doctor, he said, often consists in letting nature deal with herself, helped by suitable diet and purgatives. Contaminated atmospheres were, he thought, the cause of endemic disease and hospital and prison fevers. The best remedies were open air, free breathing, and a change of environment.

He left open the question as to whether oxidation actually took place in the lungs or in the course of circulation. Investigations already in hand on digestion and transpiration would throw further light on this.

The first paper with Seguin on transpiration[37] described an attempt to get a complete balance of the respiratory processes and ascertain the amounts of water lost through the skin and through the lungs. During the experiment Seguin was enclosed in a rubber bag, and his loss of weight was measured both with and without the bag. In addition, his absorption of oxygen and output of carbon dioxide were determined. Lavoisier could then calculate the loss of weight due to each cause on the assumption that all the oxygen absorbed formed carbon dioxide or water. If oxidation took place in the circulatory system he was uncertain whether this assumption was justified, and regarded the results as provisional.

A second paper on transpiration published after Lavoisier's death deals specially with the moisture content of air as determining comfort conditions, as to which, he said, a thermometer reading may be quite misleading.[38] The paper deals also with the purpose and design of clothing.

The trend of Lavoisier's thoughts at this time are shown by the programme drafted by him for the Academy Prize of 1794.[39] His introduction deals with the problem of nutrition, the animalization, as he calls it, of vegetable and animal food. He points out the lack of knowledge of the changes that take place in the various stages of digestion, which had led the Academy to

Antoine Laurent Lavoisier

choose an investigation into the functions of the liver and bile as the subject of the Prize. A broad treatment was suggested, including the comparative anatomy of the liver and gall bladder, the chemical nature of the bile, and the pathology of the liver. The investigations were to cover chemical researches, especially the new methods of organic analysis, Lavoisier's own invention. The programme was almost a forecast of some of the major developments in physiological chemistry in the nineteenth century.

Lavoisier was the first to submit the vital functions to an exact physico-chemical analysis. He laid the foundations of physiological chemistry. His investigations into the source of animal heat first disclosed the full significance of metabolism. If he had lived on, what contributions he might still have made! But it was not to be.

POLITICS AND THE REVOLUTION

No Frenchman saw more clearly than Lavoisier the dangers that were gathering for France—the state of her finances, her outworn social structure with all its inequalities and injustice, and the selfishness of privilege with its lack of patriotism. No one had done more than Lavoisier to forecast the reforms that had to come, and when Necker in 1787 set up the Provincial Assemblies to replace the effete local Governors he saw a great opportunity. Chosen to represent the Third Estate at Orleans he was soon its outstanding figure. 'It is Lavoisier who does everything, enlivens everything, and is everywhere', said Léonce de Lavergne. We still have his memoranda—clear, practical, convincing, and amazingly modern in outlook—on the state of agriculture, the freedom of commerce, old age insurance, savings banks, infant mortality, the need for a geological survey, and above all the reform of taxation on an equitable basis.[40] But privilege stood in the way and little was done.

The tide was running swiftly, and when the Estates General were summoned by the King in 1789 Lavoisier had fresh hopes. To his great disappointment he was only elected as a substitute deputy, but again he threw himself into the battle. He was a leading member of the Club of '89 which stood for constitutional monarchy. His memoir on the *Territorial Wealth of France* was ordered to be printed by the National Assembly.[41] It was an attempt to estimate the national income and its taxable capacity, a pioneer effort in statistics, made possible by the information Lavoisier had collected from every province through the organization of the Ferme.

Many of the reforms Lavoisier had foreseen were soon accomplished, but early in 1790 there is already an anxious note in a letter to Benjamin Franklin:[42]

> Après vous avoir entretenu de ce qui se passe dans la chimie, ce serait bien le cas de vous parler de notre révolution politique; nous la regardons comme faitte et bien faitte sans retour; . . . Les personnes modérées et qui ont conservé leur sang-froid dans cette effervescence générale, pensent que les circonstances nous ont entraînés trop loin . . . et qu'il est à craindre que l'établissement de la nouvelle constitution n'éprouve des obstacles de la part de ceux mêmes en faveur de qui elle a été faitte. . . . Nous regrettons bien dans ce moment votre éloignement de France; vous auriés été notre guide et vous nous auriés marqué les bornes qui nous n'aurions pas dû franchir.

In almost the last of Lavoisier's scientific papers there was a strangely prophetic passage:

> Faisons des voeux surtout pour que l'enthousiasme et l'exagération qui s'emparent si facilement des hommes réunis en assemblées nombreuses, pour que les passions humaines qui entraînent la multitude si souvent contre son propre intérêt, et qui comprennent dans leur tourbillon le sage et le philosophe commes les autres hommes, ne renversent pas un ouvrage entrepris dans si belles vues, et ne détruisent pas l'espérance de la patrie.

The violence Lavoisier feared came quickly. Members of the Club of '89 were marked men because of their moderation. He himself was attacked by Marat, whose ridiculous pamphlet, *Traité du Feu*, he had criticized. One by one he had to give up the offices which he had served so long and ably. In 1791 the administration of taxes was taken from the Ferme, and in 1792 he resigned from the Arsenal. Even the Academy was not safe. Lavoisier as its treasurer fought hard for its existence, but some of its own members turned against it, and it was suppressed in 1793.

Knowing the financial difficulties of France he offered his

services in various capacities, but was refused. In 1792, however, the King wished to nominate him as Minister of the Public Funds, an office Lavoisier would have occupied so gladly in happier days. He knew it was too late, that he was suspect as a member of the old regime, and no politician. He refused, saying, 'je ne suis ni jacobin, ni feuillant'.

Lavoisier found another outlet for his activities in these difficult years. In 1791 the National Assembly had set up a Consultative Committee on Arts and Crafts to advise the government on various questions, including useful inventions. Lavoisier, as usual, seems to have drafted all the reports, the most important of which was entirely his own.[43] It was on a system of public education, a subject to which he had given much thought, having started at his own expense a primary school at Villefrancoeur.

The report dealt with the general principles of education, and its national importance. It proposed a scheme covering the whole field from the primary school to the Lycée. Under his plan all children at the age of six were to enter a primary school, and at eleven were to go on to an elementary school teaching either the arts or the arts and sciences. Finally, there were to be twelve lycées giving the highest form of public education with a wide choice of subjects. Lavoisier had an eye to practical subjects like hygiene, the weather, and simple surveying. Country schools were to teach the elements of agriculture and town schools the elements of commerce.

Lavoisier is emphatic as to the importance of education in every walk of life.

Le mot *industrie* n'exprime pas toujours un emploi de forces, ni même d'adresse, il exprime le plus souvent un emploi des facultés de l'esprit. . . . Le cultivateur qui prospère le plus n'est pas toujours celui qui est physiquement le plus fort et le plus adroit; c'est celui qui est le plus intelligent. . . . Organisez l'instruction publique dans toutes ses parties; donnez du mouvement aux arts, aux sciences, à l'industrie, au commerce.

In 1791 the National Assembly had entrusted to the Academy the task of establishing a uniform system of weights and measures, a very necessary reform which resulted in the metric system. A commission was set up consisting of Borda, Lagrange,

Laplace, Monge, and Condorcet with Lavoisier as secretary and treasurer. The work was divided between various groups; Lavoisier and Haüy were to determine the density of distilled water at zero temperature. Lavoisier was also responsible for the administration of the commission, which continued after the suppression of the Academy. It was his last scientific task.[44]

Meanwhile, in the general attack on the *ancien régime*, feeling was running high against the members of the Ferme, and in 1793 they were arrested on charges of maladministration. Lavoisier asked to be allowed to continue his work on the standards, and Haüy and Borda pleaded for him in vain to the Convention, risking their own lives. All his wife's efforts, too, were unsuccessful.

When the charges against the Ferme were formulated, it was Lavoisier who drew up their defence.[45] He gave a masterly review of their financial transactions, showing that they had done their work efficiently and without undue profit to themselves. But it was the height of the Terror, the trial was a mockery of justice, and Lavoisier and twenty-seven of his colleagues were guillotined a few hours later. Lavoisier followed his father-in-law, Paulze, to the scaffold, and met death with the courage and philosophy of his generation.

Thus died this great Frenchman in his fifty-first year at the height of his achievements. The loss to France and to the world was immeasurable. The Terror had nearly run its course; within a few years came reconstruction, when under Napoleon science once more came into its own. What a contribution Lavoisier might have made to the Industrial Revolution in France, with his vision of her needs, his imaginative use of science in industry, agriculture, and hygiene, and his administrative skill and experience.

If in the fields of politics and economics he was cut off before his work came to fruition, he had foreseen the changes soon to come in industry, in social legislation, and in education. His eloquence and writings must have left their mark on the minds of those who were to remake France.

In science it was otherwise. There his work was done with almost incredible effectiveness and speed. No other revolution in scientific thought has ever been so complete, so swift, or has done so much to clear men's minds of the cobwebs of centuries.

He quickened the advance of science over its whole front. Not only chemists, but workers in physics, biology, medicine, and agriculture were given for the first time a clear view of the various forms of matter we call elements, and of the distinction between chemical and physical changes. Laviosier put new and potent weapons into their hands, just when they were needed for science to play its full part in the nineteenth century. Compare the clear-cut vision of Lavoisier's great treatise, based on quantitative measurement, on the methods of analysis and synthesis, and on the notion of the chemical equation, with the confused picture of chemistry in Macquer's textbook, with no general conception but the theory of phlogiston. It was no wonder that progress was quickened. Dalton's atomic theory, Davy's discovery of the alkali metals, his recognition of the halogens as elements, and the massive contributions of Berzelius followed swiftly, completing the picture and giving chemistry its modern form.

It was, indeed, a revolution, and I have tried to show how Lavoisier fought the battle for the New Chemistry almost single-handed, for he had no school, with many other preoccupations in one of the busiest of lives. He was fortunate in his great contemporaries, Black, Priestley, and Cavendish, who so often gave him the clue he was searching for, but he alone saw the significance of their discoveries.

What was the secret of his greatness? Lavoisier was above all things a reformer with an intense desire to improve everything he saw around him, whether it was the confused state of chemistry, the system of taxation, the lot of the peasant, the making of gunpowder or the methods of agriculture. He had the logical systematic mind of his countrymen, joined with great constructive power and a creative urge. Burning curiosity made him a born researcher. Every problem was to him a challenge for research and experiment, and for him research meant measurement, the method of modern science. He had an essentially modern outlook. He was not bound by tradition or authority. He looked his problems squarely in the face, and judged them on evidence with the shrewd common sense he showed as a man of affairs.

Lavoisier's great strength was his realism. He saw each problem in its smallest details; and no one examined them with greater care, for he was a great experimenter; but he also had great vision and saw his problems in their widest implications. He gave his visions quantitative form, in chemistry with the balance, in finance with statistics. He applied to any practical problem the same analysis and measurement that he used in the laboratory. It is often said that scientists are too remote from real life. Lavoisier is the outstanding example of the value of the laboratory mind in action. That is his lesson for us to-day.

ACKNOWLEDGEMENTS

I am greatly indebted to previous biographers of Lavoisier, and especially to the late Professor A. N. Meldrum for his scholarly researches on Lavoisier's early work. I have to thank also my secretary, Miss Josephine Wasse, for her never-failing help.

REFERENCES

[1] *Œuvres de Lavoisier*, 4, 1. Imprimerie Impériale, Paris. (1864–93).
[2] *Œuvres*, 3, 765.
[3] *Œuvres*, 5, 187.
[4] *Œuvres*, 3, 1.
[5] *Œuvres*, 3 and 4.
[6] YOUNG, ARTHUR, *Travels in France and Italy*, p. 78. Dent, London (1934).
[7] MORRIS, GOUVERNEUR, *A Diary of the French Revolution, passim*. Harrap, London (1939).
[8] *Œuvres*, 3, 106, 128.
[9] *Œuvres*, 2, 1.
[10] *Œuvres*, 2, 38.
[11] *Œuvres*, 2, 64.
[12] MELDRUM, A. N., *The eighteenth century revolution in science—the first phase*, Longmans, Green, Calcutta (1930). Lavoisier's three Notes on Combustion: 1772. *Archeion*, 14, 16 (1932).
[13] MELDRUM, A. N., *op. cit.*
[14] *Œuvres*, 2, 103.
[15] *Œuvres*, 1, 439.
[16] BERTHELOT, *La revolution chimique: Lavoisier*, p. 48. Germer Baillière, Paris (1890).
[17] BERTHELOT, op. cit., p. 246.

[18] BOYLE, ROBERT, *Essays of the Strange Subtility* ⎫
                Determinate Nature   ⎬ *of Effluviums*
                Great Efficacy   ⎭
 To which are annext New Experiments to make fire and flame ponderable: together with a Discovery of the perviousness of glass, passim. W. G. for M. Pitt, London (1673).
[19] LAVOISIER, *Observations sur la physique* (edited Abbé Rozier) **4**, 448 (1774).
[20] *Œuvres*, **2**, 105.
[21] BERTHELOT, op. cit., p. 257.
[22] *Œuvres*, **2**, 122.
[23] PRIESTLEY, J., *Experiments and observations on different kinds of air*, **2**, 320. J. Johnson, London (1774).
[24] *Œuvres*, **5**, 391.
[25] *Œuvres*, **2**, 129.
[26] *Œuvres*, **2**, 812.
[27] LAVOISIER, *Observations sur la physique*, **2**, 510 (1772).
[28] LAVOISIER, *Observations sur la physique*, **1**, 10 (1773).
[29] *Œuvres*, **2**, 283.
[30] *Œuvres*, **2**, 334.
[31] YOUNG, ARTHUR, op. cit., p. 78.
[32] *Journal Polytype*, 5 February 1786.
[33] GUYTON DE MORVEAU, LAVOISIER, BERTHOLLET, and DE FOURCROY, *Méthode de nomenclature chimique*, Paris (1787).
[34] *Œuvres*, **1**, 1.
[35] *Œuvres*, **2**, 676.
[36] *Œuvres*, **2**, 688; *Annales de Chimie*, **91**, 318 (1814).
[37] *Œuvres*, **2**, 704.
[38] LAVOISIER, *Annales de Chimie*, **90**, 5 (1814).
[39] *Œuvres*, **6**, 33.
[40] *Œuvres*, **6**, 238.
[41] *Œuvres*, **6**, 403.
[42] GRIMAUX, E. *Lavoisier*, p. 201. Germer Baillière, Paris (1899).
[43] *Œuvres*, **6**, 516.
[44] *Œuvres*, **6**, 670.
[45] *Œuvres*, **6**, 570.

3

JOHN DALTON, F.R.S.
(1766–1844)
AND THE ATOMIC THEORY†

JOHN DALTON was elected into the Royal Society in 1822 and was awarded the first Royal Medal in 1826. Dalton's name is always associated with Lavoisier's in my mind as the two men who made nineteenth century chemistry possible. There could be no greater contrast than their circumstances: Lavoisier with all the advantages of education and opportunity that wealth could give; Dalton, the son of a weaver, earned his living by teaching from the age of twelve. Meteorology was the first scientific interest of both, an interest that they maintained throughout their lives. Neither had the superb qualitative perception of Priestley or Scheele, both relied on measurement, both were striving after fundamental causes. Lavoisier worked in a great laboratory with fine instruments, Dalton with home-made apparatus, in early days in his own room, like Davy and Berzelius. The approach of both was from the physical rather than the chemical side. Lavoisier in a moment of elation at the start of his study of gases, following on his first experiments on combustion, set down his conviction that his work was destined to accomplish a revolution in chemistry and physics. Dalton gave his book the proud title—*A New System of Chemical Philosophy*. Both their claims were amply justified by subsequent events.

Lavoisier made the balance the arbiter of the chemical balance sheet, thereby destroying the myth of phlogiston and giving chemistry a rational basis for quantitative investigation. In the first decade of the nineteenth century there were many able chemists exploring the field that Lavoiser had thus laid

† Memorial Lecture at the Royal Society on 10 November 1966.

open: in Britain, Wollaston, Davy, and Thomson, in France, Berthollet, Gay-Lussac, and Proust, in Germany, Klaproth, Bucholz, and Rose, and in Sweden, Berzelius. How was it that it fell not to one of these but to Dalton, a self-taught lone worker with little experience of chemistry, to give the atomic theory the quantitative significance that made it the vehicle of chemical thought throughout the nineteenth century? This is one of the most fascinating questions that has engaged the attention of historians of chemistry for more than a century and there have been several rival answers. Now at last I think it can be answered with some certainty in spite of the destruction of Dalton's notebooks in the bombing of Manchester.

In this essay I shall consider mainly the genesis of his atomic theory, its impact on his contemporaries, and the events which made it the dominating factor in chemical theory for a hundred years.

DALTON'S EARLY LIFE

John Dalton was born on 6 September 1766 at the village of Eaglesfield near Cockermouth in Cumberland, a stronghold of the Society of Friends. His father was a hand-loom weaver with a small holding. Both his parents came of Quaker stock from which Dalton inherited his love of learning, his simple earnest character, and his sturdy independence of mind. Luckily the excellent Quaker schoolmaster in the village school taught him mathematics with a practical bent, instead of Latin, and when he left school at the age of eleven 'he had gone through a course of surveying, mensuration and navigation'. Another stroke of good fortune for Dalton was his early friendship with Elihu Robinson, a Quaker of means and influence, who was a skilful meteorologist and instrument maker. He taught Dalton in the evenings and gave him his lifelong interest in meteorology. At the age of twelve Dalton began to earn his living by opening a village school in an old barn and later in the Quakers' Meeting House. Two years later he worked for a year on his father's little farm and then joined his elder brother Jonathan in his cousin's school at Kendal which they took over in 1786 when the cousin retired. At Kendal Dalton was lucky again in meeting a remarkably able blind man, John Gough, who taught him languages and mathematics and encouraged him to start a meteorological journal in 1787, which Dalton continued until the day before his death. Dalton had a competitive nature and while he was teaching at Kendal he won many prizes for his answers to the problems set by the *Ladies'* and the *Gentlemen's Diaries*.

Dalton taught a wide range of subjects in the school and in 1787 he tried his hand at public lecturing by offering a course of twelve lectures in natural philosophy; he continued to give them in subsequent years. Then in 1793 on Gough's recommendation to the Principal of the Manchester Academy, a Dissenting College, Dalton was appointed tutor in mathematics and natural philosophy, a position he held for six years, until he decided to become a free-lance science tutor and lecturer for the rest of his life. When Dalton left Kendal in the spring of 1793 his book, *Meteorological Observations and Essays*, was already being printed in Kendal and he wrote the Preface in Manchester in September. On the first page he describes himself as Professor of Mathematics and Natural Philosophy, at the New College, Manchester, and below he quotes a line of Horace, meaning 'One can go so far, if one is not allowed to go further', suggesting that he might have reached his limit. But ten years later Dalton went much further.

The book is a remarkable work providing striking evidence of the quality of Dalton's mind, especially considering the circumstances under which it was written while he was busy teaching at a small school in Kendal, where as he says, 'not having by me all the books I could have desired I was necessarily, and perhaps luckily, forced to contemplate a good deal on the different subjects and to try such experiments as were within my reach'. During those years Dalton was pondering deeply over the significance of the data he was collecting and formulating fundamental theories of his own to account for the phenomena of the weather. Dalton's consideration of the physics of the atmosphere was the precursor of his laboratory investigations in Manchester in the years 1799 to 1803 which led him directly to his atomic theory. So his book provides important evidence of the evolution of his ideas.

His initiative in making his own barometer, thermometer, rain gauge, and hygrometer, shows Dalton's innate experimental ingenuity. With these instruments Dalton made daily observations for five years, 1788 to 1792, of the barometric pressure and the wind, of the temperature of the air, of its hygrometric state and rainfall, and also of the occurrence of snow and ice, of thunder, and in great detail of the aurora borealis. So that he had a mass of statistics to interpret. In the first part of his book he marshals them in tables of their average value in each month and in this he was no doubt helped by the standard form in which the data were given for London in the *Philosophical Transactions* for the first three years, which he quotes for comparison with his own data, and with those for Keswick made by his friend Crosthwaite.

The second part of the book contains a series of essays in which Dalton sets out his conclusions about the physical phenomena underlying his observations and their correlation. All this leads up to his simple rules for forecasting the weather based on the probability of certain sequences of events deduced from his observations. In his essay on evaporation, which is of special interest, he gives the boiling point of water under different pressures from his own measurements. He regards evaporation as a physical process and even in these early days before he knew much chemistry he is critical of the idea that the presence of water vapour in air is due to chemical combination. This was to become a major factor in his thinking in the years ahead. In these essays Dalton's originality of mind and his imaginative approach are clearly evident. All through his active life Dalton had a taste for systematic measurement but he was always striving to find an interpretation of his results. His philosophy was simple: 'The progress of philosophical knowledge is advanced by the discovery of new and important facts; but much more when these facts lead to the establishment of general laws.'

When he came to Manchester in 1793 he had for the first time access to libraries and scientific literature. Probably he was not surprised to find that some of his speculations had been anticipated and he expressed satisfaction in the confirmation of his theories by earlier workers. His theory of trade winds as due to the Earth's rotation and the heat of the tropics had been published by Hadley, and his conclusions about the causes of the variation of the barometer had been anticipated by De Luc. His pioneer experiments on dew points, however, proved that rainfall was determined by temperature and not by atmospheric pressure. Dalton had made an elaborate study of the aurora borealis and had observed both the magnetic disturbances that often accompany it and the relation of the luminous beams to the magnetic meridian. He ascribed the phenomenon jointly to the Earth's magnetism and the presence of some ferruginous or magnetic material in the upper atmosphere and he devised a mathematical theory to account for its appearance. He then found that both Celsius and Halley had anticipated his magnetic theory.

THE LITERARY AND PHILOSOPHICAL SOCIETY OF MANCHESTER

In 1794 a new vista was opened to Dalton by his election into the Literary and Philosophical Society of Manchester to which he contributed over a hundred papers on a wide variety of subjects, many of which were not published; the last was in 1844 three months before his death. His first paper soon after his election dealt with the defective colour vision of himself and his brother. This was the first scientific description of colour-blindness, to which it drew attention. It was also of value for its clinical approach although Dalton's theory of its cause was incorrect.[1]

Five years passed before Dalton's next scientific paper. Meanwhile his interest in chemistry had been stimulated by a course of lectures by Dr. Garnett in 1796, when Dalton added chemistry to his repertoire of public lectures. In 1798 he read a paper entitled, 'Essay on the Mind, its Ideas and affections; with an Appreciation of Principles to explain the Economy of Language'. Unfortunately there is no record of it. Dalton's years as a Professor were unproductive of scientific work and perhaps his decision to become a free-lance teacher in 1799 was partly in order to have more opportunities for his scientific work. Certainly the start of a new bout of experiments that led directly to his atomic theory dates from his leaving the New College, and they were done in the room which the Literary and Philosophical Society placed at his disposal for his teaching and experimental work.

In March 1799 Dalton returned to practical meteorology with a notable paper entitled, 'Experiments and Observations to determine whether the Quantity of Rain and Dew is equal to the Quantity of Water carried off by the Rivers and raised by Evaporation; with an Enquiry into the Origin of Springs'.[2] This is very pertinent to our present anxiety about water supplies. It shows Dalton's courageous and imaginative use of statistics. His objective was to discover how much water was likely to be available to replenish Britain's springs. He made estimates of the total average annual amount of precipitation in the form of rain and dew, of the run-off by rivers into the sea and the loss by evaporation. As his estimates of gains and losses were roughly equal he concluded that the springs received water by percolation during the wet seasons which they lost largely by the extraction of water from the soil by the roots of plants in the dry. Dalton with the help of his friend Thomas Hoyle carried out experiments on the percolation of rainwater through soil and the loss by run-off and evaporation in an ingenious apparatus on the lines of a modern lysimeter. His concern with the evaporation of liquids was of special importance as it led to an extensive investigation two years later which had a decisive influence on the evolution of his ideas. Once again meteorology was the stimulus to thought. The collection of the statistics embodied in this paper must have occupied Dalton for some years. It was illustrated by an interesting coloured map of the river basins. Dalton had not forgotten what he had learned during the year he had worked on his father's little farm of the value of water to plant growth.

In April 1799 Dalton read a paper on 'The Power which Fluids possess of conducting Heat' in which he controverted Rumford's view that they were nonconducting.[3] This was followed by a paper in June on 'The Colour of the Sky and the Relation betwixt solar light and that derived from combustion'[4] Dalton's mind was continually concerned with the phenomena of heat and in 1800 he carried out experiments to determine the 'Heat and Cold produced by the Mechanical Condensation of Air'.[5] In April 1800 he read a paper on 'The Expansion of Elastic Fluids by Heat' describing his experiments with a number of different gases from which he concluded 'that all elastic fluids under the same pressure expand equally by heat ... this remarkable fact that all elastic fluids expand the same quantity under the same circumstances, plainly shows that the expansion depends *solely* on heat.'

Dalton then embarked on a major programme of research the results of which he described in October, 1801, when three sessions of the Society in October were devoted to Dalton's exposition of his first theory of the constitution of mixed gases followed by accounts of his investigation of the vapour pressure of water and other liquids both in a Torricellian vacuum and in air, and also of the rate of evaporation of liquids. These were published together in 1802 with the paper on the thermal expansion of gases.[6]

Dalton's first theory of mixed gases had already been the subject of a letter to *Nicholson's Journal* in September 1801.[7] The constant composition of the atmosphere had been explained by Lavoisier, Berthollet, and others as being due to a loose chemical combination between its elements. This view was widely held at the time, for instance by Davy, who said in 1800 in his book on nitrous oxide: 'that the oxygen and nitrogen of atmospheric air exist in chemical union, appears almost demonstrative. . . . Atmospheric air then may be considered as the least intimate of the combination of nitrogen and oxygen.' Dalton, whose approach was from the physical angle, had from the first opposed this theory and held that there was no evidence of chemical combination either between oxygen and nitrogen or with the water vapour always present in different amounts in the air. In order to account for the fact that the two constituents did not separate into layers of different density, like oil and water, Dalton assumed that the particles of each gas repelled those of the same kind but were inactive towards the others so that each gas behaved as a vacuum in relation to the others. He applied the same reasoning to the vapour pressure of water at different temperatures which he had shown was independent of the presence of other gases. Having determined the vapour pressure of water over a wide range of temperatures, Dalton then investigated the relation between the rate of loss by evaporation from a free surface of water and its vapour pressure, and also the effect of air movement over the surface. Meanwhile, Dalton had been busy completing his *Elements of English Grammar* published in 1801. The Preface may well have been based on his unpublished paper of 1798, *Essay on the Mind*.

The first half of 1802 must have been occupied with experiments in preparation for a paper read in October, 'On the

Proportion of the several Gases or Elastic Fluids Constituting the Atmosphere with an Enquiry into the circumstances which distinguish the Chymical and Mechanical Absorption of Gases by Liquids'.[8] This paper, in which Dalton once again argues that the constituents of the atmosphere, including water vapour and carbonic acid, are present in a state of physical mixture was not published until 1805, so that his statement 'that the elements of oxygen may combine with a certain proportion of nitrous gas or with twice that portion but with no intermediate quantity' to form nitrous or nitric acid, was obviously added later as an example of his law of multiples. An important paper read by Dalton in January 1803, 'On the Tendency of Elastic Fluids to Diffusion through each other', which influenced his second theory of mixed gases, was also not published until 1805.[9]

In December 1802 Dalton's physical view of the solubility of gases in liquids received powerful support from his friend William Henry's investigation which showed that their solubility was directly proportional to their pressure (Henry's law).[10] Henry, originally a supporter of the chemical theory, was now converted and gave his support to Dalton in the controversy with Thomson and others.

From September 1802 we can follow the course of events from the extracts from Dalton's notebooks transcribed by Roscoe and Harden. The last four months of 1802 were occupied with experiments on the solubility of carbonic acid and other gases in water and on the reaction between lime water and carbonic acid. These experiments were continued in the early months of 1803 when the notebooks show that Dalton was busy with speculations about the mechanism, not the chemistry, of the solution of gases. There can be no doubt that it was this interest that first directed his attention to the problem of finding the relative weights of the ultimate particles of different gases. In his paper on 'The Absorption of Gases by Water and other Liquids' read on 21 October 1803,[11] Dalton says: 'The greatest difficulty, attending the mechanical hypothesis, arises from different gases obeying different laws. Why does not water admit its bulk of every gas alike? I am nearly persuaded that the circumstance depends on the weight and number of the ultimate particles of the several gases. Those whose particles being lightest and single being least absorbable and the other more as they increase in weight and complexity.' Dalton's footnote, 'Subsequent experience renders this conjecture less probable', evidently added in 1805, shows that he had lost confidence in his theory which, nevertheless, led him to the investigation that made him famous. 'An enquiry into the relative weights of the ultimate particles is, as far as I know, entirely new: I have lately been prosecuting this enquiry with remarkable success.' The determination of the relative weights of the ultimate particles had therefore become an immediate object of research. It is true that the paper was not published until 1805 and was adjusted to represent Dalton's views at that date, but the sequence of events shows that the stimulus that produced his atomic theory dates from 1803. Mr Thackray has pointed out to me that this view of the genesis of Dalton's atomic theory was first clearly stated by George Wilson in 1862 in his admirable essay on Dalton in his *Religio Chemici*. The same view was elaborated by L. K. Nash of Harvard in 1956.[12] In April Dalton was working on the elusive reaction between nitric oxide and oxygen. In May and June he returns to the expansion of liquids by heat and the force of steam. After his usual July holiday, in the first days of August he is back at nitric oxide and oxygen and the variation in their reaction under different conditions, recording on 4 August 1803: 'It appears too, that a very rapid mixture of equal parts of com. air and nitrous gas, gives 112 or 120 residium. Consequently the oxygen joins to nit. gas sometimes 1·7 to 1, and at other times, 3·4 to 1'.[13] So at the beginning of the crucial month in the history of the atomic theory Dalton had written down an example of combination in multiple proportions. How far that may have influenced the run of his mind, consciously or subconsciously, we shall never know. But it is a relevant circumstance.

DALTON'S ATOMIC THEORY

On 6 September 1803 Dalton's notebook contains for the first time tables of atomic symbols, of atomic formulae, and of the relative weights of the ultimate particles of five common elements and their compounds, compared with hydrogen as unity. Fortunately these pages were photographed by Roscoe and Harden and they are reproduced in Plates 5 and 6.

So Dalton must have been very busy with speculation,

reading, and calculation during August. He had rejected the possibility that the relative weights of the ultimate particles of gases would be proportional to their gaseous densities as he knew that steam was specifically lighter than oxygen although it contained oxygen. Hence this method was ruled out.[14] Dalton was then left with the possibility of inferring the relative weights of the ultimate particles from the proportions in which they were known to combine. Being a firm believer in the Newtonian atom he must have asked himself the question —How are the atoms likely to enter into chemical combination? He gave the simple answer—probably one to one, or if there are several compounds, one to two or three, and two to one. How often in the history of science the simple question and the simple answer have been the occasion of a flash of genius. However, there is evidence that Dalton's mind was not altogether unprepared; in his criticism of Thomson's view that nitrogen and oxygen are in some loose form of chemical combination in air, he had written in November 1802: 'Two or more heterogeneous particles may unite and become a new centre for the caloric to adhere to; but in this case the particles are no longer two but one; and oxygenous and hydrogenous would become aqueous vapour.'[15]

Dalton now proceeded to apply and test his theory by calculations from the analyses of compounds made by Lavoisier and others, including Davy. His first atomic weights were those of oxygen and nitrogen. Since he knew of only one compound of hydrogen with oxygen and with nitrogen respectively, water and ammonia, he assumed that they contained one atom of each element and from their known composition he assigned values of 5·66 and 4 to the weights of the ultimate atoms of oxygen and nitrogen compared with that of hydrogen as unity. As Dalton knew of three oxides of nitrogen, two of sulphur, and two of carbon, he was faced with the problem of assigning atomic formulae to them before he could calculate atomic weights for sulphur and carbon or check the values already found for oxygen and nitrogen. For this choice he seems to have relied on the relative densities of these gases, which were all available to him, for a clue to their relative complexity. He had also Davy's data of the composition of the oxides of nitrogen. He assigned correct atomic formulae, N_2O, NO, and NO_2 in modern style, to the oxides of nitrogen and CO_2 and CO to the oxides of carbon and to sulphuric and sulphurous acids, SO_2 and SO. These gave him values of 17 and 4·5 for the atomic weights of sulphur and carbon from the known analyses of sulphuric and carbonic acids. The available values for the lower oxides were too discrepant and inaccurate to help him.

Dalton's next step is to test his atomic weights and page 246 of the notebook shows him examining the data for the composition of the acids of nitrogen given by Cavendish and Lavoisier to see if they agreed, using his atomic formulae as the link between them. On pages 247 and 248 he makes a direct test of his atomic weights assuming that nitric oxide (nit. gas) contains one atom of each element:

'Ult. atom of nit. gas should therefore weigh 2·42 azote
Ult. atom of oxygen should therefore weigh 1·42 azote
According to this 1 oxygen will want 1·7 nitrous.'

It is hard to believe that he was not struck by the coincidence of this number with the results he had written down on 4 August for the combination of oxygen and nitrous gas. They are not strictly comparable as the latter refer to volume and the former to weight, but Dalton may have forgotten this as the units are not specified in his notebook. We know that he attached great importance to the oxides of nitrogen as a test of this theory from Mr Thackray's reconstruction of the war-damaged manuscript of a talk that Dalton gave to the Literary and Philosophical Society in 1830:

'I had frequent . . . with Mr (afterwards Sir Humphry) Davy . . . discussed the merits of the atomic . . .

(bottom of page burnt away)
(page 2)
Nitrous oxide, nitrous Gas and nitric (nitrous) acid, which Sir Humphrey had so ably investigated in his Researches then recently published. I was the more happy in this as his results formed some of the most excellent exemplifications of the principles. The *heavy* nitrous oxide consisting of two atoms of azote and one of oxygen, the *light* nitrous gas consisting of one atom azote and one of oxygen, and the *heavy* nitrous acid gas, consisting of one atom of azote and two of oxygen, seemed well adapted to afford an illustration of the principles. From the

John Dalton

observations of Sir Humphry however the speculation appeared to him rather more ingenious than important.'[16]

Page 250 contains Dalton's correct calculation of the composition of the three oxides of nitrogen on the basis of his atomic weights and the values of Cavendish, Lavoisier and Davy for the composition of nitric acid (NO_2), the mean of which is near to his own calculation:

Page 250

On the position of the other pages if nitric acid and other compounds of azote be such

	A.	Ox.
then Nitric Oxide	58·6	41·4
Nitrous gas	41·4	58·6
and Nitric acid	26	74
Cavendish	27·7	72·3
Lavoisier	$20\frac{1}{2}$	$79\frac{1}{2}$
Davy	$29\frac{1}{2}$	$70\frac{1}{2}$

It is curious that he does not compare Davy's values for the other oxides with his own as he certainly knew them. Perhaps he did and failed to record it. On 20 December 1803, when he was at the Royal Institution he made his comparison and found a rough agreement helped by an error in his arthimetic.[17]

Summing up the events of August 1803 one can say that, luckily for chemistry, Dalton's faith in his new theory was not shaken, as it might have been if his critical acumen had been more exacting and the data available to him less discrepant. He shows his confidence by proceeding at once to apply his theory by calculating from his atomic weights and the known specific gravity of the gases the relative diameters of the ultimate particles of a number of both elementary and compound gases. The values are given in a table on page 258 dated 19 September 1803. Two pages later he arranges the gases in the order of their atomic weights (he says specific gravities but the values he uses are the relative weights of the ultimate atoms) in order to test his idea, which had initiated the whole enquiry into the weights of ultimate particles, namely that these may determine their relative solubilities. He found that the order was roughly the same but he seems later to have been doubtful about a causal relationship.

In October 1803 Dalton made a few experiments on the

reaction between nitric oxide and oxygen, but the bulk of his work was devoted to the gas analysis of the composition of various gaseous compounds of carbon including seventy experiments on ether vapour leading to formulae for ether and alcohol. On 21 October he read his paper on the Absorption of Gases to the Manchester Society. It was not published until 1805, and represents his views at that time rather than those he held in 1803. It contained the first published tables of his relative weights of ultimate particles, the values being those he adopted in 1805, and it included values for carburetted hydrogen and olefiant gas that were not determined until 1804.

In December 1803 Dalton went to London to lecture at the Royal Institution when he discussed his atomic theory with Davy. The early months of 1804 were directed to experiments to find the temperature of the maximum density of water. In March Dalton is calculating the atomic weights of metals assuming simple formulae for the oxides. In May and June he is working on the diffusion of gases and in August, coming back refreshed from a holiday when he examined air from Helvellyn, he returns to the eudiometric analysis of gaseous compounds of carbon and on 24 August he made his decisive experiments on marsh gas and olefiant gas proving that their composition obeyed his rule of multiples.[18] This must have given Dalton immense satisfaction, as he always had greater confidence in his own measurements than in those of others, and for the first time he had made a crucial test of his theory with his own hands.

Two days later Thomas Thomson spent a day or two in Manchester and he records that 'he was much with Mr Dalton. At that time he explained to me his notions respecting the composition of bodies. I wrote down at the time the opinion that he offered.' Dalton had no doubt emphasized the importance of the evidence he had just discovered to support his theory and he evidently gave Thomson the impression 'that the atomic theory first occurred to him during his investigations of olefiant gas and carburetted hydrogen gases at that time imperfectly understood, and the constitution of which was first fully developed by Mr Dalton himself.' So this was the erroneous story of the genesis of Dalton's theory which Thomson gave to the chemical world in the third edition of his *System of*

John Dalton

Chemistry in 1807,[19] and repeated in his *History of Chemistry* in 1830,[20] which was generally accepted for nearly a century.

The final months of 1804 were devoted to various experiments on heat including the rate of cooling of bodies. In 1805 Dalton went to London to buy apparatus and in the summer he gave a course of lectures in Manchester, doing little laboratory work, but in September he draws up 'more correct tables of the sp.gr. &c of various gases' using the latest values of the relative weights of their ultimate particles and their densities to calculate afresh their relative diameters. The differences he finds in these led to his second theory of mixed gases.[21] I agree with Thackray in assigning this to 1805 not 1804. Dalton was always a little doubtful about his assumption that atoms only repelled those of their own kind, and he now assumes that repulsion among atoms is a general characteristic. The phenomenon of diffusion he now explains by the differences in the size of the particles which would prevent their being stacked in a stable formation and would allow them to jostle past one another with mutual diffusion. After some experiments on the effect on air of breathing and combustion, leading to a paper on this subject in March 1806, Dalton spends the next months speculating on the relationship between heat and his atomic particles and their specific heats. In the later months of 1806 his interest in the chemical aspects of his theory is evidently increasing and he begins to see its many applications. By 1807 Dalton was anxious to give his own account of his new theory and he arranged to give courses of lectures in Edinburgh and Glasgow in March and April.[22] Much of the rest of the year was devoted to the preparation of part I of *New System of Chemical Philosophy* which appeared in 1808. It represents the peak of Dalton's achievements.

THE NEW SYSTEM OF CHEMICAL PHILOSOPHY

The keen competitive spirit in Dalton that had so often won prizes for the answers to questions in the *Gentlemen's* and *Ladies' Diaries* was still alive and he dashed off incisive letters to *Nicholson's Journal* replying to criticism of his ideas even when they came from his old benefactor, John Gough. He also arranged for the publication of William Henry's letter telling him that he was a convert to his physical conception of the solubility of gases. It must have been a great satisfaction to Dalton that his papers had attracted the notice of scientists on the Continent, even when they disagreed with him. So the time had now come when he wanted to publicize his views more widely and show that he was capable of breaking a lance with De Luc, de Saussure, and even Berthollet himself. In the preface to part I of his *New System* he mentions 1803 as the culminating year when 'he was gradually led to these primary Laws, which seemed to obtain with regard to heat and to chemical combinations'. Part I is an autobiographical account of the physical theories of heat and of the characteristics of liquids and gases that Dalton had developed in the years since he had begun to keep a meteorological record. In this way it resembles Lavoisier's *Traité élémentaire* and the tables in which Dalton collects his systematic measurements are reminiscent of the prototype collection of physico-chemical data that Lavoisier annexed to the *Traité*. In spite of Berzelius's criticism of Dalton's accuracy, even in the 1833 edition of his *Lehrbuch* he quotes Dalton's table of the boiling points of solutions of sulphuric acid. In part I Dalton is the professional physicist describing his own experiments on which his theories were based, and he might have claimed the independent discovery of three physical generalizations. Anyone who wants to understand the run of Dalton's mind, based on Newton's atoms, must read his part I carefully. The first two hundred pages deal entirely with his physical outlook and his opposition from his early years to the ideas of Lavoisier, Berthollet, Davy, and others that the atmosphere is a loose chemical compound of oxygen and nitrogen. Dalton's atomic concept was quite clear, either two particles retain their identity or if they unite they become one. He cannot contemplate any intermediate stage. His work on the vapour pressure of liquids, following de Saussure, convinced him that the vapour pressure was independent of the pressure of another gas, and consequently physical in nature. In one interesting biographical passage he relates his surprise that while lime water quickly extracted all the carbonic acid from air, water alone did not do so even on long standing as he had expected. This must have been a critical moment when he realized the difference between chemical combination and physical solubility and by 1808 Dalton had realized, as his title shows, that his exploration of the physical aspects of meteorology had in the end led him to propound a new theory which explained the quantitative basis of chemical combination. So the final chapter of part I, entitled 'On Chemical Synthesis', contained a précis of his atomic theory with a plate of his atomic symbols representing elements and compounds and a list of atomic weights. He left the application of his theory to part II and gave no clue as to its genesis either by deduction or by *a priori* assumption, so his readers are left guessing as to this.

Two years later came part II when Dalton had carried out a number of simple chemical experiments to satisfy himself. In the preface he says: 'Having been in my progress so often misled by taking for granted the result of others, I have determined to write as little as possible, but what I can attest by my own experience. On this account, the following work will be found to contain more original facts and experiments, than any other of its size, on the elementary principles of chemistry.' In part II we see Dalton as the naïve amateur chemist repeating many well-known experiments and adding some of his own to demonstrate the applications of his atomic theory and the means of determining atomic weights. He ends with a table of the relative weights of many single and compound atoms, with pictures of the atomic formulae of fifty compounds, with plates showing the different diameters of the single and compound atoms, and graphs of the boiling points of solutions of nitric and muriatic acids and of ammonia. His final table contains his evaluation of the composition of seventeen gaseous compounds. Among his experiments are some showing his keen observation and the pleasure he derived from his discovery of the photochemical reaction between oxymuriatic acid and hydrogen, which ceased 'when a cloud obscured the Sun'. But it would be no exaggeration to say that nobody did less justice to Dalton's theory than Dalton himself. Berzelius was justified in his criticism of the crude inaccurate experiments by which Dalton justified to himself his theory that muriatic acid, oxymuriatic acid, and fluoric acid are all different oxides of hydrogen. Here Dalton was certainly shaping nature to agree with his hypothesis.

Fortunately for chemistry the future of his theory depended

not on Dalton himself but on its impact on his contemporaries to which I shall now turn.

THE IMPACT OF DALTON'S THEORY ON HIS CONTEMPORARIES

Davy was the first to hear of it when Dalton was lecturing at the Royal Institution in December 1803. Davy was not impressed. As we have seen, to Sir Humphry 'the speculation appeared rather more ingenious than important'. However, Davy recognized the importance of the law of multiple proportions and in 1811 wrote to Berzelius: 'In the two last Bakerian Lectures, those for 1809 and 1810, I have embraced the doctrine of definite proportions', and it certainly clarified his thinking. Berthollet shared Davy's scepticism about Dalton's atoms and in his introduction to the French translation of Thomson's *System of Chemistry* said that the atomic theory was not only 'speculative' but 'seductive'.

Thomas Thomson was the next to be told of Dalton's theory and we know from a letter from Thomson to Dalton that the account he gave of it in his *System of Chemistry* in 1807 was based on his talk with Dalton in 1804. Thomson was an enthusiastic supporter of Dalton's ideas and he used his symbolic formulae several times in his textbook. A year later in a paper read to the Royal Society on Oxalic Acid he produced valuable evidence confirming the law of multiples by his analysis of the neutral and acid oxalates.[23]

A fortnight later Wollaston gave Dalton his authoritative support in a paper on Subacid and Superacid Salts, which shows that he may have been on the point of discovering the law of multiples himself.[24] After referring to Thomson's evidence for this law he says: 'As I had observed the same law to prevail in various other instances of subacid and superacid salts, I thought it not unlikely that this law might obtain generally in such compounds, and it was my design to have pursued the subject with the hope of discovering the cause, to which so singular a relation might be ascribed. But since the publication of Mr. Dalton's theory of chemical combination, as explained and illustrated by Dr. Thomson, the enquiry which I had designed appears to be superfluous, as all the facts that I had observed are but particular instances of the more general

John Dalton

observations of Mr. Dalton, that in all cases the simple elements of bodies are disposed to unite atom to atom singly, or, if either is in excess, it exceeds by some simple multiple of the number of its atoms.'

Wollaston then describes a number of simple experiments comparing the composition of acid and normal salts, which anyone could repeat to convince himself of the evidence for the law of multiples. At the end of the paper Wollaston makes his well-known forecast of the need to consider the spatial arrangement of the primary particles 'before any theory of chemical combination can be rendered complete'.

But Wollaston like Davy was sceptical of Dalton's atoms and his paper on a Synoptic Scale of Chemical Equivalents in 1813 was to become the source of much confusion later.[25] Wollaston made oxygen the basis of his table of equivalents which were calculated in a rather arbitrary way and were in fact a mixture of atomic weights and equivalents. Wollaston realized their value in chemical calculations and devised a slide-rule graduated for that purpose.

Two anonymous papers appeared in the *Annals of Philosophy* in 1815 and 1816, known to have been written by William Prout, pointing out that many atomic weights referred to hydrogen as unity were close to whole numbers.[26] On the assumption that they were really whole numbers Prout calculated the specific gravity of a number of gases and vapours and showed that these were close to the observed values. On this evidence he based his hypothesis that the elements were condensed forms of hydrogen which represented the πρωτη ὑλε of the Greek philosophers. Prout's hypothesis became the will-o'-the-wisp that attracted so much speculation and research for a century until the mystery of the elusive approximation of many atomic weights to whole numbers was finally explained by Aston in 1919 with his mass spectrograph.

In France, when Gay-Lussac in 1809 wrote his classic paper on the combination of gaseous substances with one another,[27] he had read the account of Dalton's theory given by Thomson and also Dalton's *New System*. Although his own law of combination by simple volume ratios seems complementary to Dalton's multiple proportions, Gay-Lussac, influenced by Berthollet's opinion, was sceptical about the views of both Prout and

Dalton that combination could only occur in simple proportions by weight. Dalton was equally sceptical about Gay-Lussac as his own less accurate measurements did not support Gay-Lussac's simple law.

It was left to the clear logic of Avogadro in Italy,[28] uninfluenced by any preconceived assumptions, to draw the logical conclusions from the work of Gay-Lussac in the light of Dalton's theory and to show the inconsistencies arising from Dalton's arbitrary assumptions, although in some cases he agreed with his conclusions. Avogadro had also given careful attention to Davy's calculations of proportional numbers. It must remain a matter of conjecture as to how far the run of Avogadro's mind was influenced by Dalton's theory, which he developed so skilfully in the light of Gay-Lussac's Law, explaining the difficulties which Dalton and Gay-Lussac had seen in the densities of steam and carbonic oxide being less than that of oxygen which they both contain. Avogadro's clear distinction between the molécule élémentaire, i.e. the atom, the molécule constituante, the molecule of an element, and the molécule intégrante, the molecule of a compound, and his logical application of the hypothesis that equal volumes contain equal numbers of molécules, constituantes or integrantes, led him to the diatomic conception of the molecules of the gaseous elements, an idea so hostile to the current atomic theory that it took fifty years and bitter fights before it won general acceptance. It was strongly opposed by Berzelius, being incompatible with his electro-positive and -negative polar atoms, his explanation of chemical affinity. In two later papers in 1814 and 1821 Avogadro calculated a number of atomic and molecular weights correctly. But although the theory soon had independent confirmation from Ampère, it attracted little support until much later.

Like Avogadro, Ampère was drawn to molecular theory by Gay-Lussac's discoveries.[29] His main objective, however, was to show that various geometrical assemblies of molecules would explain both the different crystal systems and Haüy's Law of Rational Indices. At Berthollet's request he sent him an outline of his theory for publication in the *Annales de Chimie* in 1814. In his crystal structures Ampère assumed that the molecules of elements are tetrahedral, containing four atoms (in modern

John Dalton

terms). Having reached the same conclusion, independently of Avogadro, that equal volumes of gases contain equal numbers of molecules under the same physical conditions, Ampère showed as a necessary consequence from Gay-Lussac's discoveries that the molecules of nitrogen, oxygen and hydrogen must be divisible into two parts, each of them according to him containing two atoms. He still considered chlorine as an oxide of an unknown element. Unlike Avogadro, he did not consider any other chemical implications of his theory and he published no further papers on the subject. However, in spite of his erroneous assumption of a tetrahedral molecule, it was Ampère rather than Avogadro that Gerhardt, Laurent, and Gaudin quoted in their papers on molecular theory. Ampère does not mention Dalton in his paper.

BERZELIUS AND THE ATOMIC THEORY

I come now to the great Swedish chemist, Berzelius. It was Berzelius who gave Dalton's atomic theory the form that made it the chemist's shorthand, and became the main vehicle of chemical thought for a hundred years. Berzelius first knew of Dalton's theory when he read Wollaston's paper on subacid and superacid salts in *Nicholson's Journal* early in 1809. By then he had already spent two years in refining the methods of analysis of metallic oxides and salts in order to test the validity of Richter's generalization, based on the continuance of neutrality when two neutral salts interact. Berzelius found that his result gave ample confirmation of Dalton's law of multiples, in fact it was only his conventional method of expressing his results in percentages that had prevented him from discovering it for himself. Not having seen Dalton's *New System* he proceeded cautiously. In the first of several papers published in 1811,[30] he said he would abstain from theorizing and would let the results speak for themselves. This they did, revealing many simple ratios of the oxygen in different oxides of the same metal and in the basic and acid constituents of salts, following the dualistic conception of Lavoisier. In 1812 Dalton sent him a copy of his *New System*. Berzelius wrote to Gautier de Claubry in April: 'I have never had a present that gave me so much pleasure at first. As yet I have only been able to look through it hastily, but I will not conceal that I am surprised to see how the author has deceived my hopes. Incorrect in even the mathematical part, e.g. the determination of the maximum density of water, in the chemical section he allows himself deviations from the truth that are indeed surprising, one sees how he seeks everywhere to shape nature to his hypothesis'.[31] It is easy to understand Berzelius's impatience with Dalton's crude volumetric measurements when he was striving to attain an accuracy of a few thousandths in his gravimetric work. However, he realized the great significance of Dalton's theory and the flash of genius behind it. Later he wrote to David Brewster: 'If one takes away from Dalton everything but the atomic idea, that will make his name immortal. What remains is only a collection of badly executed experiments and of mathematical conclusions that are quickly refuted, although by their mathematical form they begin by carrying conviction.'[32] Berzelius had a vein of sarcasm with which at an unfortunate moment he was to touch Davy on the raw. Thirty years later he wrote in retrospect in calmer vein: 'I soon convinced myself by new experiments that Dalton's values lacked the accuracy that was needed for the practical use of the theory. I recognized that if this new light was to illuminate the whole science that it was necessary to determine, with the greatest possible accuracy, the atomic weights of as many elements as possible, particularly those of the most common elements, in order to evaluate the proportions in which the compound atoms combine, as for example in salts, the composition of which I had been investigating for some time. Without this no daylight could follow the rosy dawn. This was therefore the most important objective of chemical research, to which I devoted myself unceasingly. At intervals I investigated afresh many of the most important atomic weights using improved techniques. After ten years' work I was able to publish in 1818 tables containing the composition of two thousand substances, calculated from the results of my own atomic weight determinations.'[33]

Berzelius was a systematist of the school of Linnaeus and with his encyclopaedic knowledge he looked at chemistry as a whole. In 1812 he published a new system of chemical nomenclature advocating the use of Latin names as a chemists' lingua franca.[34] This did not catch on but it paid a dividend in 1814 when he introduced the use of the first one or two letters of the

Latin name of each element to represent an atom in an atomic formula.[35] In this year also he published his first tables of atomic weights or volume weights as he called them as he based them on proportions by which gaseous elements combined by volume.

In 1813 Berzelius had reformed the nomenclature of the Swedish pharmacopoeia[36] and in 1814 he published a new system of mineralogy based on his electrochemical theory and on atomic composition.[37] In 1815 he proved that organic compounds, like inorganic, obey Dalton's law of multiples although in a more complex form.[38] By 1818 Berzelius had determined single-handed the atomic weights of almost every known element and in 1819 he issued his comprehensive tables of the weights of simple and compound atoms.[39]

In spite of his great knowledge and quick eye that had discovered three new elements, Berzelius was conservative, particularly with regard to the postulates of his system, and his choice of some of them was unfortunate. He assumed that in all inorganic compounds one element must be present as a single atom. Dalton rightly criticized this assumption,[40] but Berzelius was stubborn, until in 1819 the discovery of the law of isomorphism and of the equality of atomic heats, both the direct fruits of Dalton's theory, compelled him to introduce formulae, like Fe_2O_3 and to halve all his atomic weights for metals, bringing most of them near to their present values. He also clung to Lavoisier's theory that all acids and bases contain oxygen and in his atomic weight tables of 1818 muriaticum and nitricum still figure as the mythical elements of which chlorine and nitrogen were the oxides. This view also he abandoned about 1820, so that by 1826 most of his atomic weights were near to their present values.[41] Berzelius referred his atomic weights to oxygen as 100, since he had determined most of them by direct comparison with oxygen. Many chemists had adhered to Dalton's standard of hydrogen as unity as this gave smaller and more convenient values. So in his 1826 Tables Berzelius had two sets of values, referred to oxygen and hydrogen respectively. He then introduced, purely from motives of convenience, a change in his formulae that was to have unfortunate repercussions. As the halving of the molecular weights of the metals involved doubling the number of metallic atoms in the formulae, Berzelius said that it would be more convenient and less liable to error to draw a horizontal stroke through the symbol to denote a double atom.[42] He applied this generally so that the formula of water became H̄ = H_2O. Thus he introduced barred atoms without attaching to them any theoretical significance. Unfortunately in his 1835 Tables instead of referring his Daltonian atomic weights to H = 1 he referred them to H̄ = 2, consequently they ceased to be comparable with those on the oxygen standard and became equivalents.[43] Some chemists had followed Wollaston's lead in preferring equivalents to atomic weights and the publication of Gmelin's large *Handbook* in 1843 using equivalents had a considerable influence, so it was no wonder that there were rival sets of formulae, all, however, using the Berzelian atomic notation with different numerical values.

Towards the end of the 1820s many of the younger chemists, particularly in France and Germany, directed their researches to the field of organic chemistry, the compounds of carbon, which yielded a rich harvest. Following the example of Scheele and Chevreul they concentrated their attention on complex natural products, plant acids and bases, oils and fats and compounds of animal origin like uric acid. Few of the simple carbon compounds were then known. Gradually a large number of these complex substances were separated, purified, identified, and analysed. It was here that the Berzelian formulae played an indispensable part. Without these shorthand expressions of the composition of each of these complex substances, it would have been an impossible task to unravel their relationships and their reactions. The interpretation of the formulae led to rival views as to the constitution of these carbon compounds, the stimulus to further investigations, thanks to which by 1843 the index of Liebig's *Organic Chemistry* contained some two thousand substances each with its own formula.[44]

During these years Berzelius had the first school of research chemists in Stockholm, long before Liebig's at Giessen, with Mitscherlich, Wöhler, and the two Roses working with him and using his atomic weights. At the same time he published each year the first *Annual Reports on the Progress of Chemistry*.[45] These for nearly twenty years made him the accepted arbiter, backed by his encyclopaedic knowledge, the prestige of his great

Lehrbuch and the generalizations he embodied in the words isomer, polymer, and catalysis, which he coined and used in the precise sense that we attach to them today.

CLOUDS ON THE HORIZON

In the late 1830s however clouds began to gather around Berzelius. The investigation of the anomalous vapour densities of mercury, phosphorus, and sulphur by Dumas, confirmed by Mitscherlich, undermined confidence in Berzelius's volume atomic weights and in 1836 Dumas in his *Philosophie Chimique* wrote: 'If I were a dictator, I should banish the word atom from our science, convinced that it goes far beyond our experience, and in chemistry we must never outstrip experience.'[46]

Then the discovery by Dumas in 1834 that chlorine could replace hydrogen, forming compounds like trichloracetic acid, very similar to its parent substance, followed by Laurent's discovery that the chlornaphthalenes are isomorphous with naphthalene, posed an insoluble problem for Berzelius. He had explained chemical affinity by his polar atoms. It was impossible for him to admit that an electronegative chlorine atom could replace electropositive hydrogen in an atomic structure. So he tortured the formulae to try to show that trichloracetic acid had a different structure to its parent, until he took refuge in Gerhardt's copulae and admitted substitution there. Equally it was impossible for him to admit the combination of two hydrogen atoms in a molecule, so Avogadro got some sarcastic comments in the *Annual Reports*.

Berzelius still held to Lavoiser's conception of the dualistic nature of salts with the assumption that both acid and base were present. This had received its death-blow from Davy's chlorine and his theory of hydracids. Nevertheless, the strength of the Berzelian tradition was shown by Liebig as late as 1843, when after describing Davy's theory of acids as compounds of hydrogen replaceable by metals, he adds, 'However, in describing these compounds I shall use the ordinary convention', and this was dualistic.[47]

It was Gerhardt who led the attack on the Berzelian dualism and electrochemical theory and in 1845 he started a new journal to get a hearing for his own and Laurent's views and to reply to the criticisms of 'the illustrious chemist of the North'.[48] One of my prized possessions is Alexander Williamson's set of Gerhardt's *Comptes Rendus*. Luckily he didn't bind the last volume as one part was missing and so the covers were preserved, which would have been destroyed in binding. On two of them there is the pathetic advertisement of the little private laboratory which Gerhardt and Laurent opened to eke out their exiguous existence. It was the most bitter period of chemical controversy and Gerhardt and Laurent were persecuted for their views. It's a tragic story ended by Laurent dying from sheer want. It was Laurent, when defeated by Balard for a professorship, who made the bitter comment, 'Balard, who was discovered by bromine'.

However, in spite of all the controversies that centred on Berzelius in these stormy years, chemists all continued to use the Berzelian shorthand for their formulae. I often wonder whether Berzelius ever recalled his comment on Haüy's criticism when he was defending the young Mitscherlich's discovery of isomorphism in his *Annual Report* of 1822: 'Haüy's contention is thus completely untenable. But one cannot expect that a scientist, towards the end of a life full of honour, should surrender, without resistance, without any attempt at defence, a position which he regarded as the most important of his discoveries. That is perhaps more than one has a right to expect from any man.'[49]

EVOLUTION OF THE CHEMICAL EQUATION

Here I must turn aside for a moment to consider the evolution of the chemical equation, an important subject that seems to have escaped the attention of historians of chemistry. It is appropriate that the first chemical equation with the sign of equality was written in 1787 by Lavoisier, who had made the balance the arbiter of the chemical balance sheet. It dealt with one of the oldest chemical reactions known to mankind, which has had such a stimulating effect on his thought and action down the ages:

$$\text{grape juice} = \text{alcohol} + \text{carbonic acid}$$

Lavoisier omitted the oxygen on the left-hand side.[50]

Today the chemical equation is such an obvious sequel to the use of atomic formulae that it might have been expected to follow quickly on their introduction, but that was not so. I cannot remember that Berzelius ever used an equation. Dalton on one occasion did so. The first paper in which I have found the habitual use of the equation, showing that it was the author's normal approach, is in Laurent's doctoral thesis of 1837.[51] Liebig in his *Textbook* fo 1841 uses it very occasionally, Gerhardt in his translation more freely, but in his own *Textbook* of 1844 the equation is common form.[52] My guess is that this is one of the many innovations concerning the atomic theory that we owe to those constructive iconoclasts, Gerhardt and Laurent.

THE PERIOD 1850–1860

During the 1840s the use of different systems of atomic weights or equivalents by individuals did not retard progress if each system was used consistently while the investigation of new compounds gradually covered the whole field. But by 1850 when attention was directed more and more to the simpler compounds containing only a few carbon atoms, the use of correct values of molecular and atomic weights became essential for the gradual emergence of the concepts of valency and atomic structure. So by 1850 the correct determination of molecular weights had become a crucial issue. There was however a fundamental difference of opinion among chemists as to the evidence they would accept. Gerhardt, Laurent, Gaudin, and later Cannizzaro were strenuous supporters of Avogadro's use of vapour densities, but the majority of chemists would accept only chemical evidence such as Williamson's classic paper in which he established the molecular formulae of the alcohols and ethers.

I have a paper in the handwriting of William Odling, Gerhardt's 'l'ami Odling', one of the main protagonists of the 1850s. It is part of the notes he used in a lecture I persuaded him to give in 1899 to the Junior Scientific Club in Oxford on 'Chemical theories under discussion about the year 1850. Some personal reminiscences'. Luckily it got mixed up with the secretary's papers and so it was preserved. This is what he says: 'Means for determination of unit weight. Conclusions from gas or vapour densities and other physical considerations. Deduction of molecular or unit weights scarcely accepted by physicists and not at all by chemists. Necessity for determination of unit weights on chemical grounds for establishment of concordance of results with those based on physical grounds.' Here he is expressing the current view of most chemists, although Williamson's results were in accordance with the deductions from Avogadro's law.

The confusion resulting from the rival systems of equivalents and atomic weights can be seen in the four different formulae in use for water, H_2O, H_2O_2, HO and HO, while nineteen formulae for such a simple substance as acetic acid covered a whole page of Kekulé's *Textbook* in 1858.

Cannizzaro and Kekulé independently took steps to end this chaos. Cannizzaro at Genoa was determined that his students should be saved from the confused thinking of the period so he gave them a series of lectures on the historical development of the atomic theory in which by the logical use of Avogadro's Law, eliminating all the preconceived notions which had bedevilled theory, he arrived at the values of atomic and molecular weights that we use today and showed further that they were in accordance with the evidence from isomorphism and atomic heats, and from the recent research on the dissociation of certain compounds. A précis of his views was published in 1858 in the Italian journal *Nuovo Cimento*,[53] but it attracted no attention.

Kekulé in 1859 persuaded Weltzien and Wurtz to join him in organizing the first international chemical conference at Karlsruhe in 1860 to see if some agreement could be reached.[54] One hundred and forty of the leading chemists duly met at Karlsruhe on 3 September 1860, including Dumas, Cannizzaro, and Mendeleev. Unfortunately Kekulé was implacably opposed to accepting physical evidence, and the conference was occupied largely by long arguments between him and Cannizzaro on this issue and eventually it broke up having reached no conclusion, although accepting the retrograde view of Dumas that they should return to Berzelius's atomic weights and barred atoms. The irony of this was that Dumas had done more than anyone to discredit Berzelius, and the purpose of his proposal was to continue his vendetta against Gerhardt who had died in 1856. But this was not the end. As they were departing, Pavesi distributed a few copies of Cannizzaro's précis and

PLATE 4

JOHN DALTON
From the portrait by R. R. Faulkner

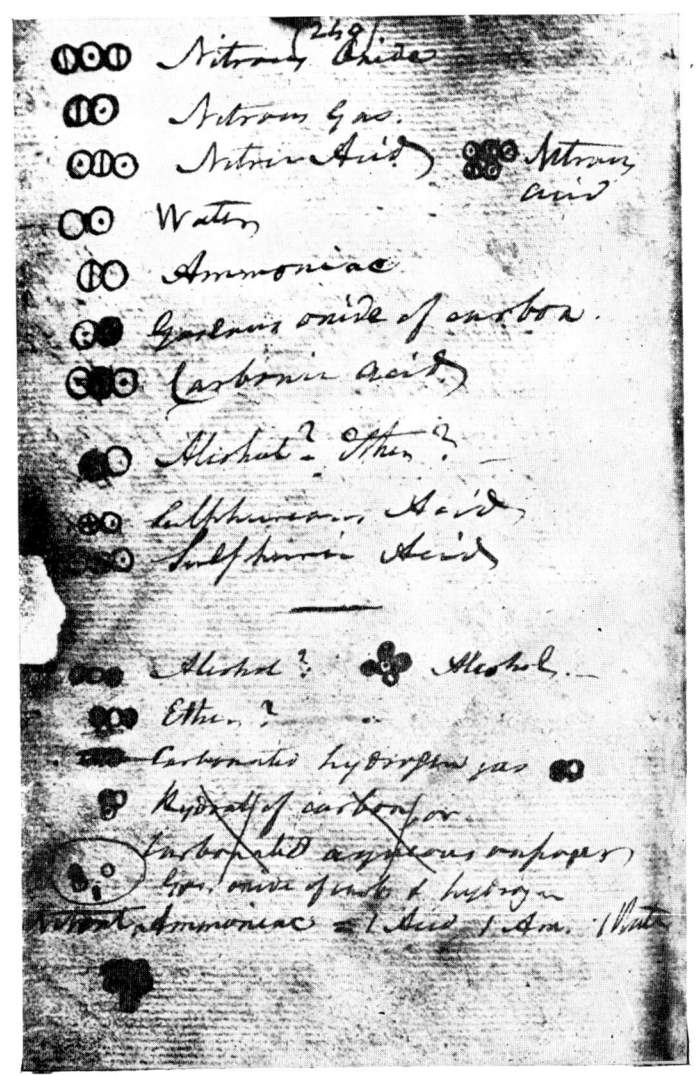

Page from Dalton's note-book

①⊙①	Nitrous Oxide.
①⊙	Nitrous Gas.
⊙①⊙	Nitric Acid.

⊙①⊙
①⊙   Nitrous Acid.

○⊙	Water.
①○	Ammoniac.
⊙●	Gaseous oxide of carbon.
⊙●⊙	Carbonic acid.
●○	Alcohol? Ether?
⊕⊙	Sulphureous Acid.
⊙⊕⊙	Sulphuric Acid.

●○●   Alcohol?

○
●⊙●   Alcohol.
○

●○●   Ether?

Carbonated hydrogen gas ●○

●
⊙   Gas. oxide of carb. & hydrogen.
⊙

Nitrat ammoniac = 1 Acid 1 Am. 1 Water.
⊙①⊙
○⊙○
 ①

Page from Dalton's note-book

Observations
on the ultimate particles of Bodies
& their combinations.

Sept. 6 1803
Characters of Elements—
○ Hydrogen.
☉ Oxygen.
⊕ Azote.
● Carbone, pure charcoal.
⊕ Sulphur.

Ult at. Hydrogen	1
Oxygen	5·66
Azot	4
Carbon (charcoal)	4·5
Water	6·66
Ammonia	5
Nitrous gas	9·66
Nitrous oxide	13·66
Nitric acid	15·32
Sulphur	17
Sulphureous acid	22·66
Sulphuric acid	28·32
Carbonic Acid	15·8
Oxide of carbone	10·2

HUMPHRY DAVY
From the portrait by Sir Thomas Lawrence

A page from Davy's note-book

both Lothar Meyer and Mendeleev got one. Lothar Meyer read it on his way home and as he said 'The scales fell from my eyes'.[55] He recognized its logic and the clarity it brought to chemists' thinking. It was the main theme of his *Moderne Theorien*, written two years later, which was a major influence in getting general agreement, except in France, to our modern systems of atomic and molecular weights. This, together with Couper's and Kekulé's carbon bonds and Kekulé's benzene formula, made possible the great development of organic chemistry in the later decades of the nineteenth century. Mendeleev wrote later: 'The decisive moment in the development of my theory of the periodic law was in 1860, at the conference of chemists in Karlsruhe, in which I took part, and at which I heard the views of the Italian chemist, S. Cannizzaro. I regard him as my immediate predecessor, because it was the atomic weights which he found which gave me the necessary reference material for my work. I noted immediately that the modifications he proposed to the atomic weights, introduced a new pattern into Dumas's groupings, and it was then that I was struck by the essential idea of a possible periodicity in the properties of the elements on increase of atomic weight.[56] So the Karlsruhe Conference, thanks to Cannizzaro, in the end achieved its object. Mendeleev's Periodic Law was the most far-reaching product of the atomic theory.

STEREO-ISOMERISM

In 1874 came most striking evidence of the potentialities of the atomic theory when van't Hoff and Le Bel independently pointed out that a tetrahedral grouping of four dissimilar groupings around carbon atoms would account for their optical rotatory power and for the existence of the dextrogyral and laevogyral forms of tartaric and other acids discovered by Pasteur in 1847.[57] There is an interesting story of the events leading up to this which I have never seen properly told. In 1830 Berzelius discovered that tartaric acid and paratartaric acid (later known as racemic acid) had identical atomic compositions, although the former was optically active and the latter inactive. This with other examples led Berzelius to his generalization of isomerism. Mitscherlich had at that time undertaken to write a crystallographic Supplement to Berzelius's

Lehrbuch. Berzelius wrote at once to him asking him to measure some salts of both acids to see if they were identical.[58] A month later Mitscherlich replied that the salts of the two acids he had measured had different crystalline forms and as usual he was full of excuses for the delay in finishing the Supplement.[59] In 1831 he wrote to say that the sodium ammonium salts of the two acids were, however, identical.[60] Meanwhile, Mitscherlich dilly-dallied and the Supplement was never finished. Eleven years later in 1842 Mitscherlich published a short note in the *Berichte* of the Prussian Academy describing the identity of the crystals of the two salts although one is optically active in solution and the other inactive.[61] Biot with his great interest in optical rotation noticed this and arranged with Mitscherlich to present a note to the Académie in 1844, to which he attached a note from himself confirming Mitscherlich's statement about the optical properties of solutions of the two salts, but differing from his view that their atomic structures are identical.[62]

Now it so happened that the young Pasteur was at that time a student in the Ecole Normale at Paris and that he was dreaming about the relationship between the dextrogyral and laevogyral forms of crystals and their optical rotatory powers, in some cases in the crystalline state, in others in solution. Mitscherlich's statement was a direct contradiction of the theory he had formed and so he repeated the experiments and made his classic discovery that sodium ammonium racemate gives a mixture of the crystals of the dextrogyral and laevogyral tartrates.[63] Fortunately the racemic double salt was unstable at the temperature at which crystallization was carried out. Accident has often played a part in scientific progress. What were the odds against Mitscherlich's measurements catching the eye of the young Pasteur as a student in 1844, if they had been published in Berzelius's Supplement in 1831?

DALTON'S LATER YEARS

The final triumph of Dalton's atomic theory came in this century when X-ray diffraction proved that the atomic formulae, which chemists had derived from macroscopic evidence, represent very closely the actual assemblies of atoms in the molecule. For a century the Newtonian–Dalton atom had served chemists well and by 1920 the structure of chemical molecules

John Dalton

that it had revealed began to receive further elucidation by the more intimate conceptions of Rutherford and Bohr of the structure of the atom itself.

That, very briefly, is the story of some of the major achievements of Dalton's atomic theory in the nineteenth century, and now we must return to Dalton, where we left him after the publication of the second part of his *New System* in 1810. Dalton was then forty-four; he had shot his bolt. Part I of the *New System* was the peak of his achievement. His long devoted study of his first love, meteorology, had led him to an intensive investigation of the physical properties of liquids and gases in the light of the Newtonian atom. In 1803 this had culminated in his realization of the quantitative mechanism underlying chemical combination, his unique contribution to chemical theory. Dalton was far less happy when he tried to apply his own theory to chemistry. He lacked both the techniques and the judgement that he had shown in physical investigations. Some of his later papers were refused publication and the declining interest in his pronouncements can be gauged by the fact that the first part of volume two of his *New System*, presented to the Royal Society's library by the Manchester Literary and Philosophical Society in 1827, remained on the shelves with the pages uncut until August 1966. I had some hesitation in disturbing its virginity. But Dalton was by now a famous man. His Royal Medal in 1826 was followed by his election as a foreign member of the Academies of Berlin, Munich, and Moscow, and in 1830, after Davy's death, Dalton was elected in his place as one of the eight foreign Associates of the French Academy, of which he had been a corresponding member since 1816. In 1832 he received a D.C.L. at Oxford, and a year later he was granted a Royal pension. Dalton continued to work in the laboratory and to teach. Of his many pupils, the last, James Prescott Joule, was the most famous. Although Dalton only taught him for two years, owing to an attack of paralysis in 1837, it was Dalton who encouraged Joule to start research in a laboratory in his father's house. Dalton continued his meteorological record to the end of his life, the last entry being made on the day before his peaceful death on 27 July 1844. A fortnight later Manchester honoured her most famous citizen by a public funeral.

By the irony of fate it fell to Davy, the first sceptic of Dalton's atomic theory, to read his citation when he presented him with the first Royal Medal. Davy's Presidential Addresses set the model that future Presidents were to follow.

After reciting the pre-Daltonian history of the atomic theory, mentioning the works of the two Higgins and of Richter, Davy goes on: 'Mr Dalton, as far as can be ascertained, was not acquainted with any of these publications, at least he never refers to them: and whoever will consider the ingenious and independent turn of his mind, and the original tone prevailing in all his views and speculations will hardly accuse him of wilful plagiarism. But let the merit of discovery be bestowed wherever it is due: and Mr Dalton will still be pre-eminent in the history of the theory of definite proportions. He first laid down clearly and numerically the doctrine of multiples, and endeavoured to express by single numbers, the weights of the bodies believed to be elementary. . . . Mr Dalton's permanent reputation will rest upon his having discovered a single principle, universally applicable to the facts of chemistry—in fixing the proportions in which bodies combine, and thus laying the foundation for future labours, respecting the sublime and transcendental laws of the science of corpuscular motion. His merits, in this respect resemble those of Kepler in astronomy.'[64] These words were spoken before Dalton's theory had began to exert its full influence on the progress of the science. Of its unique value in later years there could be no better judges than Liebig and Cannizzaro, who had been in the heat of the battle, and here is what they said. Liebig in a letter to William Henry, Dalton's biographer, written in 1853 says: 'You wish to have from me, my view of the hypothesis of Dalton, and of its influence on the development of chemistry. This is a difficult problem; for we, who stand in the presence of the science as now constituted, can scarcely conceive how it would have developed itself without this hypothesis. All our ideas are so interwoven with the Daltonian Theory, that we cannot transpose ourselves into the time, when it did not exist. Dalton's atomic theory was a product of the age. . . . Chemistry received in the atomic theory, a fundamental view; which overruled and governed all other theoretical views, to which the ideas of the age respecting chemical forces, affinity, cohesion referred themselves; it was the bond which bound together all other views. In this lies the extraordinary service which this theory rendered to science.'[65] Cannizzaro in his Faraday Lecture of 1872 re-echoed Liebig's words: 'Chemists who have adopted this language of the atomic theory have not been able to change its meaning: hence it has resulted that the atomic theory has become more and more interwoven with the warp of chemical science, and can no longer be separated from it without tearing the whole tissue.'[66]

So we can leave John Dalton with the words of Davy, Liebig and Cannizzaro as his epitaph, for as Berzelius said, his name is immortal.

ACKNOWLEDGEMENTS

I am greatly indebted to Mr Arnold Thackray for his scholarly record and analysis of Dalton's own words which show so clearly the run of his mind.

I owe to Dr Maurice Crosland the reference to Berthollet's criticism of the atomic theory in the French translation of Thomson's *System of Chemistry*.

REFERENCES

[1] *Mem. Manchester Lit. Phil. Soc.* **5**, 28–45 (1798).
[2] Ref. (1), 5 part 2, 346 (1802).
[3] Ref. (1) 373.
[4] Not published.
[5] Ref. (1), 515.
[6] Ref. (1), **5**, 535–602 (1802).
[7] *Nicholson's Journal* (first series), **5**, 244 (1802).
[8] Ref. (1) (second series), **1**, 244 (1805).
[9] Ref. (1) (second series), **1**, 259 (1805).
[10] HENRY, W., *Phil. Trans.* **93**, 29, 274 (1803).
[11] Ref. (1) (second series) **1**, 286 (1805).
[12] NASH, L. K., *Isis*, **47**, 101–116 (1956).
[13] ROSCOE AND HARDEN. *A new view of the origin of Dalton's atomic theory*, p. 38 (1896).
[14] Ref. (13), 27.
[15] Ref. (7), **3**, 271 (1802).
[16] THACKRAY, A. W., Ref. (1), **108**, 2.
[17] Ref. (13), 44.
[18] Ref. (13), 63.
[19] THOMSON, T., *A system of chemistry*, **3**, 424 (1807).

[20] THOMSON, T., *History of chemistry*, **2**, 289 (1831).
[21] Ref. (13), 16–17.
[22] Ref. (13), 141–143.
[23] *Phil. Trans.* **98**, 63–95 (1808).
[24] *Phil. Trans.* **98**, 96–102 (1808).
[25] *Phil. Trans.* **104**, 1–22 (1814).
[26] *Ann. Phil.* **6**, 321, 472 (1815); **7**, 111 (1815).
[27] *Mémoires de la Sociéte d'Arcueil*, **2**, 207–34 (1809).
[28] *J. de Physique*, **73**, 58–76 (1811).
[29] *Annls. Chim.* **90**, 45–86 (1816).
[30] *Annls. Phys.* **37**, 249 (1811).
[31] *Jac. Berzelius Bref.* (Uppsala, 1919), **7**, 104.
[32] Ref. (31), **7**, 7.
[33] BERZELIUS, J. J., *Lehrbuch der Chemie* (5th ed, 1843), **3**, 1161.
[34] *Kgl. Vetenskaps—Academien handlingar* (Stockholm, 1812), **33**, 28.
[35] *Ann. Phil.* **3**, 51–62, 93–106, 244–57, 353–64 (1814).
[36] BERZELIUS, J. J., *Svenska lakaresallskapets handlingar*, I, 11–100 (1812).
[37] BERZELIUS, J. J., *An attempt to establish a pure scientific system of mineralogy by the application of the electro-chemical theory and the chemical proportions* (London, 1814).
[38] *Ann. Phil.* **4**, 409; (1814) **5**, 93–101, 174–84, 260–75 (1815).
[39] BERZELIUS, J. J., *Essai sur la Théorie des Proportions Chimiques* (Paris, 1819).
[40] *Ann. Phil.* **3**, 174 (1814).
[41] BERZELIUS, J. J., *Lehrbuch der Chemie* (edition 1826), **7**, 397, **8**, 1 and 177.
[42] BERZELIUS, J. J., *Jahresbericht Fortschr. phys. Wiss. Tubingen*, **7**, 72 (1828).
[43] BERZELIUS, J. J., *Théorie des Proportions Chimiques*, Paris (1835).
[44] LIEBIG, J., *Handbuch der Organischen Chemie*, Heidelberg (1843).
[45] BERZELIUS, J. J., *Jahresbericht über die Fortschritte der physischen Wissenschaften* 1822–47.
[46] DUMAS, *Leçons sur la Philosophie Chimique*, p. 290. Paris, (1836).
[47] Ref. (44), 6.
[48] *Comptes Rendus Mensuels des Travaux Chimiques*, per Mm. Aug. Laurent et Charles Gerhardt, Paris (1846).
[49] Ref. (45), **1**, 71 (1822).
[50] LAVOISIER, A., *Traité Élémentaire de Chimie*, **1**, 141.
[51] LAURENT, A., *Recherches Diverses de Chimie Organique*, Paris (1837)
[52] GERHARDT, C., *Précis de Chimie Organique* (1844).
[53] *Nuovo Cimento*, **7**, 321–66 (1858).

54 HARTLEY, H., Stanislao Cannizzaro and the first International Chemical Conference at Karlsruhe in 1860, *Notes and Records of the Royal Society*, **21**, 56 (1966).
55 Ostwald's Klassiker Nr. 90, *Abriss eines Lehrganges der Theoretischen Chemie vorgetragen an der K. Universität Genua* von Prof. S. Cannizzaro. Herausgegeben von Lothar Meyer. Verlag Engelmann, Leipzig (1891).
56 FIGUROVSKY, N. A., Dmitrii Ivanovich Mendeleev. *Izv. Akad. Nauk, SSSR*, pp. 44–51. Moscow (1961).
57 VAN'T HOFF, *Chemisch Constitutie van Organische Verbindungen* (Utrecht, 1874), LE BEL. *Bull. Soc. Chim.* **22**, 337 (1874).
 The saga of van't Hoff's doctorate is still told at Utrecht. His classic pamphlet on Stereochemistry had met such criticism from the professors that he submitted for his doctoral thesis an innocuous investigation of cyanoacetic and malonic acids. How van't Hoff must have chuckled to himself. I met him in 1899 so I know that he could chuckle.
58 *Jac. Berzelius Bref*, **6**, part 1, 157. Uppsala (1932).
59 Ref. (58), 161.
60 Ref. (58), 173.
61 *Ber. K. Preüss. Akad. Wiss. Berlin*, 1842, 244.
62 *C.R. Acad. Sci. Paris*, **19**, 719 (1844).
63 PASTEUR, L., *C.R. Acad. Sci. Paris*, **26**, 535 (1848).
 PASTEUR, L., *Leçons de Chimie professées en 1860*, Ostwald's Klassiker (1891).
64 DAVY, H. *Collected Works*, **7**, part 1, 93.
65 HENRY, W. C., *Memoirs of the life and scientific researches of John Dalton*, p. 134. London (1854).
66 CANNIZZARO, S., *J. Chem. Soc.* **2** (ser. 10), 941 (1872).

4

SIR HUMPHRY DAVY, Bt., P.R.S.
(1778–1829)†

A VISITOR to the Royal Society's apartments in Burlington House passed on the staircase, first the bust of our Royal Founder and Patron by Nollekens and then the romantic portrait by Lawrence of the most romantic of our Presidents, Humphry Davy, whose name, next to that of Newton, is the most widely remembered of them all. And rightly so, for it was his discovery of the principle of the safety-lamp that made possible the great expansion of coal mining, the basis of our national wealth. It removed so largely the risk of explosions and saved untold human agony and suffering. The invention of the lamp, based as it was on a fortnight's brilliant work in the laboratory, was typical of Davy's genius for making a decisive experiment.

Davy's scientific work was done mainly in the first two decades of the nineteenth century. It was a formative period in which chemistry having shaken off the leading strings of medicine and pharmacy was emerging as an independent science. It was a time of great activity in chemical research. Every year saw some new line of attack and a new harvest of chemical discoveries. Lavoisier had established the balance as the arbiter of the chemical balance-sheet but the laws of chemical composition only came to light in the new century. Dalton's atomic weights had opened a new chapter of the atomic theory and by 1820 Berzelius's chemical formulae were fast becoming the chemists' shorthand for the communication of their results.

There were still many doubts in chemists' minds, doubts about the nature of light and heat and doubts about the elements. The idea behind phlogiston still lingered on, the idea that attributed common causes to analogous effects. That was

† The Wilkins Lecture to the Royal Society on 5 March 1953.

Humphry Davy

the basis of Lavoisier's fallacy that oxygen was common to all acids—his most troublesome legacy to chemistry. These were the years when these doubts were gradually resolved and chemistry was taking on its modern form and throwing fresh light on all the related sciences. It was acquiring, too, a new importance in the public eye, for chemical industry had become an important factor in the industrial revolution in Great Britain. The contributions of Berthollet and Monge to the Napoleonic economy had their influence elsewhere.

That is the background against which Davy's work must be seen and judged. He was fortunate in the problems of this formative period, when there were many discoveries to be made in which his genius for qualitative experiment and his objective outlook found full scope.

Davy was born at Penzance on 17 December 1778, the eldest of five children. His father, a small landowner and wood carver, came of old Cornish stock. There must have been a strain of intellectual quality in the family, as Davy's brother John was distinguished as a military surgeon and chemist, his cousin Edmund was Professor of Chemistry in Dublin and Sir Humphry Rolleston was John Davy's grandson. Davy was a precocious infant, and he was popular with the boys at the local grammar school for his gift of telling stories and for helping them to translate Latin into English verse.

There is a revealing passage in one of his notebooks. 'After reading a few books, I was seized with the desire to narrate, to gratify the passions of my youthful auditors. I gradually began to invent and form stories of my own. Perhaps this passion has produced all my originality. I never loved to imitate, but always to invent: this has been the case in all the sciences I have studied.' Davy's lively imagination was the mainspring of his work. All through his life he loved to have an audience, and nobody knew better how to hold one.

Davy grew up as a countryman, a keen observer of nature, with a deep appreciation of natural beauty and a passion for fishing and shooting which lasted all his life. Most of his boyhood was spent in Penzance living with a friend, Mr Tobin, as his family had left to live on a small farm at Varfell. Mr Tobin had a good library and Davy was a quick reader with wide interests. He left school when he was fifteen with a fair knowledge of Latin and Greek and for a year continued his own education, learning French, until he was apprenticed to a Penzance surgeon, Mr Bingham Borlase, with the idea that he should later go to Edinburgh and qualify as a physician. There is no record of the progress of his medical studies but he was popular and efficient in administering first aid. In his leisure hours he had ambitious plans for his education, first in mathematics, then metaphysics and history, and his youthful essays show him grappling with some of the most difficult problems of philosophy and religion.

In 1797 Davy started reading natural philosophy, and in December when he was just nineteen his interests turned to chemistry and he began to read Lavoisier's *Traité* and Nicholson's *Chemical Dictionary*, where he quickly found his metier. Within a few weeks without any formal training he started research in his bedroom, according to his brother, with homemade apparatus. But two of his earliest experiments were made with an air-pump which hardly falls within this category. Robert Hunt, a Fellow of the Royal Society, who wrote Davy's life for the *Dictionary of National Biography*, speaks of his early friendship with a Quaker, Robert Dunkin, a saddler, a man of original mind and of the most varied acquirements. Dunkin had constructed a number of instruments and models with which he is said to have instructed Davy in the rudiments of science. According to Hunt, Davy took Dunkin to the Larigin river on a winter's day to show him that rubbing two plates of ice together developed sufficient heat to melt them.† So Dunkin may have helped in these early experiments, though John Davy makes no mention of him.

It was characteristic of the wide sweep of Davy's mind and of his confidence in himself, that he chose for his first investigation the nature of light and heat. He knew that friction and percussion produce heat and he was anxious to decide between the rival theories, that heat consists of an elastic fluid called caloric or is due to a peculiar motion of the particles of matter. After showing that ice could be melted by friction, he made a crucial

† Professor Andrade (*Nature, Lond.* **135**, 359 (1935)) and Professor McKie (*ibid.*, p. 878) have drawn attention to the difficulties and physical impossibilities of some of these early experiments, without impugning the veracity of Davy's account of what he thought he had discovered.

Humphry Davy

experiment by melting wax in a vacuum by the heat generated by the friction of a wheel, driven by clockwork, rubbing against a piece of metal. The experiment was done at 32°F and the apparatus was mounted on a block of ice with a channel containing a small amount of water which remained unaltered during the experiment. Hence, argued Davy, no caloric could have entered the system and 'heat . . . which is the cause of our peculiar sensations of hot and cold, may be defined as a peculiar motion probably a vibration of the corpuscles of bodies, tending to separate them. . . . The phenomena of repulsion are not dependent on a peculiar elastic fluid for their existence, or caloric does not exist.' However faulty the evidence, the ingenuity of the experiment was remarkable for a beginner. At the time Davy did not know of Rumford's experiments on the heat generated in the boring of cannon which had been published a year earlier.

Davy's attack on the nature of light was less happy. His first experiment was to snap a flintlock in a vacuum and in carbon dioxide, when he saw none of the sparks that are produced in air. Hence he concluded that light is not a modification or an effect of heat. Repetition of the experiment in oxygen gave brilliant sparks. In his *Traité* Lavoisier quotes Berthollet's view that light has a great affinity for oxygen and can combine with it, 'et qu'elle contribue avec le calorique à le constituer dans l'état de gaz'. Davy had no doubt read this and thought he could improve on Lavoisier. He had found that lead oxide produced oxygen 'exposed to the light of a burning glass or even of a candle' but not 'when heated as much as possible included from light', that is at a lower temperature, which Davy forgot. This experiment he regarded as a decisive proof that gaseous oxygen was a combination with light, which he therefore christened phosoxygen. His view was confirmed by the light emitted in combustions in oxygen.

From this false start, Davy embarked on a long series of experiments on the physical and chemical effects of light, on phosphorescence, on oxygen and its compounds, on respiration, and on the effects of light and various gases on the growth of and and marine plants. He was led on, I suspect, by an unusual purple passage in Lavoisier's *Traité* in which he praises the powers of light. 'Sans la lumière la nature étoit sans vie, elle étoit morte et inanimée: un Dieu bienfaisant, en apportant la lumière, a répandre sur la surface de la terre l'organisation, le sentiment, et la pensée.' Echoes of this passage recur often in Davy's essay.

Davy's youthful experiments contain many faulty observations, and they led to much wild speculation, which has been sternly criticized. Nevertheless, his investigations included some remarkable achievements. Davy was, I believe, the first to show that carbon dioxide is present in venous blood, a fact that was still in dispute among physiologists many years later. He showed that fish can only live in water containing dissolved oxygen and that they expire carbon dioxide. He found that marine plants absorb carbon dioxide in sunlight and emit oxygen, thus serving the same function as land plants. He tried to test the truth of Lavoisier's assumption that oxygen is present in muriatic acid, by passing phosphorus vapour over heated calcium muriate to see if any phosphoric acid is formed. He got a negative result but this experiment foreshadowed his greatest contribution to chemistry ten years later.

In spite of its errors and wishful thinking the paper describing Davy's investigations was an astonishing feat for a youth with no scientific training. His breadth of treatment and his experimental insight are remarkable. Its interest today is in showing Davy's innate gift for experiment, not only in ingenuity of method but in the logical sequence of attack on a broad front. The experience he had gained in those first twelve months was soon to find another outlet.

When Beddoes read the paper with its references to the theories of a celebrated medical philosopher, Dr Beddoes, he no doubt saw in it good publicity for his Pneumatic Institute and published it in a volume of collected essays early in 1779 with a commendation from himself. The paper nautrally met with a hostile reception and Davy was quick to see his mistakes. A year later in *Nicholson's Journal* he recanted. In his notebook he wrote: 'I began the pursuit of chemistry by speculation and theories ... I was perhaps wrong in publishing with such haste a new theory of chemistry. My mind was ardent and enthusiastic. I believed I had discovered the truth. Since that time my knowledge is increased—since that time I have become more sceptical.' And Davy remained the sceptic, the experimenta

Humphry Davy

inductive philosopher, with doubts about the atomic theory, with which perhaps later he infected Faraday. It was unfortunate, for the atomic theory was the road which the progress of chemistry inevitably had to follow.

Just when Davy's interests were turning to chemistry he met by chance, most luckily, Gregory Watt, the gifted son of James Watt, and Davies Gilbert (then Giddy) who was destined to succeed him as President of the Royal Society. Gilbert noticed him swinging on the half-gate of Borlase's house. He asked who he was and was told 'Son of Davy the carver. He is interested in chemistry.' Gilbert gave him the run of his library and befriended him in many ways. Gregory Watt was sent to Cornwall as he was suffering from tuberculosis, and Watt's Cornish agent arranged for him to lodge with Davy's mother. The two young men with common interests quickly became friends. Davy must have benefited by Watt's scientific training and thanks to him was soon in correspondence with Dr Beddoes, who had just started his Pneumatic Institute at Bristol.

THE PNEUMATIC INSTITUTE AT BRISTOL

Thomas Beddoes was an erratic genius, a pupil of Joseph Black. He had been obliged to resign from his Professorship of Chemistry at Oxford owing to his radical politics, and then, with the financial support of friends like Thomas Wedgwood, he had started a clinic at Bristol for the study of the curative properties of the newly discovered gases for various diseases. Beddoes needed an assistant and at Gilbert's suggestion he offered the post to Davy. He had been impressed by Davy's letters and his training with Borlase must have given him experience in handling patients. So in October 1798 Davy migrated to Bristol. There he found ample opportunities for his experiments and stimulating society. Mrs. Beddoes, the witty and charming younger sister of Maria Edgeworth, was the centre of a literary group, then living in Bristol, and through her Davy met Coleridge and Southey with whom he soon became intimate. They enjoyed the liveliness and ubiquity of Davy's mind and his enthusiasm for his experiments. Davy must have been a most attractive young man and their letters show the admiration and affection they felt for him. They had, too, a common interest in poetry—Southey once said of Davy 'He had all the elements of poetry, he only wanted the art'. The truth is that although Davy had great facility in writing verse, he lacked the art and today his poetry seems to us rather poor stuff. But at that time both Coleridge and Wordsworth were interested in science and they valued Davy's literary judgement.

In July 1800 Wordsworth is writing to Davy from Grasmere asking him to look over the proofs of *Lyrical Ballads* written by Coleridge and himself. 'You would greatly oblige me . . . by correcting anything you find amiss in the punctuation, a business at which I am ashamed to say I am no adept. . . . In future I mean to send the MSS. to Biggs and Cottle with a request that along with the proof-sheets they may be sent to you.' In October Coleridge writes to Davy that 'Wordsworth is fearful you have been much teased by the printers on his account'.

In the following February Coleridge writes for Davy's advice about a plan of Calvert's, 'an idle, good-hearted and ingenious man, to build a small laboratory and become a fellow student with me and Wordsworth in chemistry. . . . Wordsworth has not yet decided but he is strongly inclined . . . he feels it more necessary for him to have some intellectual pursuit less clearly connected with deep passion than poetry.' The laboratory was not built but Wordsworth was soon writing that well-known passage which he added to the preface of the second edition of *Lyrical Ballads*: 'Poetry is the first and last of all knowledge—it is as immortal as the heart of man. If the labours of Men of Science should ever create any material revolution, direct or indirect in our condition and in the impressions which we habitually receive, the poet will sleep then no more than at present, but he will be ready to follow the steps of the Man of Science. The remotest discoveries of the chemist, the botanist, the mineralogist will be as proper objects of the poet's art as any upon which it can be employed.'

Cottle, the Bristol publisher, once asked Coleridge how Davy compared with the clever young men he met in London. 'Why, Davy can eat them all,' was the reply, 'there is an energy, an elasticity, in his mind which enable him to seize on and analyse all questions, pushing them to their legitimate consequences. Every subject in Davy's mind has the principle of vitality. Living thoughts spring up like turf under his feet.'

Davy was only nineteen when he went to Bristol, and the friendship and society of Coleridge and Southey must have been a great intellectual stimulus to him and an immense help in the days so soon to come, when he had to face a London audience.

NITROUS OXIDE

In March 1798, soon after he began to study chemistry, Davy luckily came across Dr Mitchill's *Theory of Contagion*.† Dr Mitchill, an American physician, had published a small duodecimo volume in 1795 seeking to prove that the 'gaseous oxyde of azote' was the principle of contagion and if breathed was followed by immediate extinction of life. This, he said, would account for the sudden deaths of those struck by the contagion of the plague. The theory aroused Davy's suspicion, which he quickly put to the test by exposing wounds and the bodies of animals to the gas and by breathing it himself mixed with air without, as he says, 'any remarkable effects'. He informed Dr Beddoes of his experiments which he had to discontinue until he could make larger quantities of the gas. These experiments were now renewed and on 11 April 1799 Davy discovered that pure nitrous oxide could be breathed without danger and experienced the sensations with which most of us are so familiar. This led him to an investigation the results of which were published in a separate volume of 580 pages just over a year later. In the introduction Davy makes his apology— 'Early experience taught me the folly of hasty generalization'.

The book contains a record of a long series of investigations of the effects of breathing nitrous oxide and other gases, both on animals and on himself. It also contains his experimental survey of the compounds of nitrogen with oxygen and hydrogen which fills more than half the volume. In studying the chemistry of nitrous oxide he found many discrepancies and gaps in existing knowledge and so he made a systematic study of the oxides and acids of nitrogen and ammonia, determining their densities and composition and making a very thorough examination of their chemical properties. It was his first purely chemical research, done at the age of twenty-one, and although it was largely a

† Davy had probably read the reprint of this book in Part V of the pamphlets written by Beddoes and James Watt on the *Medicinal Uses and Production of Factitious Airs*.

repetition of isolated experiments made by others, it showed again Davy's gift for experiment, his grasp of a wide field, and the logical sequence of his work. He does not hesitate to challenge, with justification, some of the results of Lavoisier, von Humboldt, and Vauquelin, and his own quantitative results are remarkably accurate considering his apparatus and his lack of experience. It was a great feat to have covered so much ground single-handed in a year in the intervals of his duties in the Institute and his physiological experiments. It was only made possible by the systematic nature of his attack and his careful documentation of the work of others. The adverse criticisms of the previous essays were soon forgotten on the publication of his book which at once made him an authority in this field.

The research had started with Davy's interest in the physiological effects of nitrous oxide and he examined the results of breathing this gas by warm and cold-blooded animals, fish, and insects. He was obviously greatly interested at this time in the physiology of respiration and compared the effects of nitrous oxide with those of other gases. He also made a number of experiments with animal and human blood *in vitro*. Finding that animals lived too short a time for accurate measurements he began to experiment on himself. These experiments continued for many months and in some of them he ran grave risks. Trials with water gas and nitric oxide might easily have proved fatal. Davy's objective was to discover the chemical changes that take place during the respiration of different gases. By measuring the volumes of gas breathed and expired he found that a considerable proportion of nitrous oxide was absorbed in his lungs. He saw that any results he might deduce from analyses of the inspired and expired air would depend on a knowledge of the residual capacity of the respiratory organ after expiration. This he set out to determine by breathing pure hydrogen for half a minute, having first shown that it is not absorbed in the lungs, and analysing the expired gases. From their content of nitrogen, oxygen, and carbon dioxide he calculated that the exhausted capacity of his lungs was 41 in^3 at 98°F, thus anticipating the work of Bréhant sixty years later.

He was now able to determine the amounts of gases absorbed and liberated in the lungs during respiration. From his results he concluded that oxygen was absorbed by the blood and oxidation took place during circulation, with the formation of water and carbon dioxide which were liberated from the venous blood through the moist coats of the vessels. Nitrous oxide absorbed by the blood did not react with it but displaced a small amount of dissolved nitrogen. Davy at this time thought that air was a compound of oxygen and nitrogen and this handicapped him in interpreting his results.

The last section of the book described the sensations of a number of observers including Coleridge and Southey who breathed nitrous oxide, usually failing to share in Davy's enthusiasm for 'the pleasure-producing gas'. He records in the paper that 'a desire to breathe the gas is always awakened in me by the sight of a person breathing, or even by that of an airbag or an air-holder', so he must have had many wishful moments in the Pneumatic Institute.

When he was cutting a wisdom tooth Davy discovered the power of the gas to remove intense physical pain, and he forecast its use with advantage during surgical operations in which no great effusion of blood takes place. It might also be useful, he thought, in cases of extreme debility and he mentions its use in paralytic affections. His final word is that the application of pneumatic chemistry in medicine is an art in infancy 'which must be nourished by facts, strengthened by exercise and cautiously directed in the application of her powers by rational scepticism'.

Davy was only twenty-two when the book was published. It won for him at once a high reputation as a chemist. It is interesting to speculate how much Davy's opportunity owed to Dr Mitchill's theory which would never have survived examination by a scrupulous referee.

Meanwhile, the Pneumatic Institute was flourishing with Davy's stimulating presence and quickness of wit. At one time he records that they had eighty out-patients. The treatment of palsy by nitrous oxide seems to have been one of their successes but apparently faith healing played its part. Coleridge told Dr Paris the story of a patient supposed to be suffering from paralysis who was selected to test the healing powers of nitrous oxide. Beddoes had first impressed on him the certainty of success. Davy put a small thermometer under his tongue to take his temperature; no sooner had the 'patient felt the thermometer between his teeth than he concluded that the talisman was in full operation, and in a burst of enthusiasm declared that he already experienced the effects of its benign influence throughout his whole body. . . . Davy cast an intelligent glance at Mr Coleridge and desired the patient to renew his visit on the following day when the same ceremony was performed.' After a fortnight he was dismissed as cured and Davy had to confess to Dr Beddoes the delusion he had practised.

When the work on nitrous oxide was drawing to a close Davy's keen eye was quick to see the significance of Volta's discovery of the galvanic pile and he was soon at work examining the chemical effects produced by the electric current. In his first note in September 1800 he shows that oxygen and hydrogen are produced in electrolysis 'nearly in the proportions required to form water'. Hence he was led to suppose that the constituent parts of compound bodies 'might be separately extricated' and obtained distinct from each other. He was surprised that from a strong solution of caustic potash he got only oxygen and hydrogen showing that no decomposition of potash had taken place. Davy's interest was now thoroughly aroused and in each of the next four months he had a short paper in *Nicholson's Journal* describing investigation of the powers of piles made with various pairs of his metals and connecting fluids from which he concluded that galvanic action depends on the oxidation of one of the metals.

Those were Davy's last experiments in Bristol. Count Rumford was looking for someone to replace Dr Garnett, the Professor of Chemistry at the Royal Institution. Davy was recommended to him by several friends. And so in March 1801 Davy became assistant lecturer in chemistry and experimenter to the Royal Institution with the prospect of becoming Professor in the future.

THE ROYAL INSTITUTION

The Royal Institution was founded in March 1799 as the result of a proposal made by Count Rumford and Mr Thomas Bernard, both of them practical philanthropists, to form a Public Institution by private subscription for diffusing the knowledge and facilitating the general introduction of useful

Humphry Davy

mechanical inventions and improvements. A collection of working models, a laboratory and lectures covering a wide range of the applications of science were the first aims of the Managers. The house in Albemarle Street was quickly bought and reconstructed and Rumford lived there at first to superintend the activities of the Institution. A passage in a letter from him to Sir Joseph Banks, the Chairman, about the Romsey Public Kitchen shows the practical interests of the founder. 'I shall bring with me to Town a very clever Bricklayer of his neighbourhood who is desirous of completing his education under my auspicies at the Royal Institution. Our Roaster here has been publickly tried and the meat roasted in it was unanimously preferred to meat roasted in the common. I cannot finish the letter without communicating to you a very interesting and if I am not mistaken a most important discovery. The Process of cooking meat called *boiling* may be performed with a degree of heat considerably below that of boiling water,—and the meat so cooked is uncommonly savory and high flavoured. A Piece of the toughest neck Beef was made very tender in about three hours. . . . I shall not fail to push these enquiries to the utmost.'

Dr Garnett was appointed as the first Professor of Natural Philosophy and Chemistry and his first lectures in 1800 were crowded with 'distinction and fashion'. Even before Davy's day coachmen had to set down and take up with horses' heads towards Grafton Street. In March 1800 a scientific committee of the Managers had been set up 'to examine the syllabuses of the professor of natural philosophy and chemistry to the end that no false doctrine might be taught at the Institution and to superintend all the new philosophical experiments that might be made in the house of the Institution'. Garnett was soon in trouble for not submitting his syllabus and in his second course 'his spirited way of lecturing changed into languor and hesitation' and he was asked to resign. Meanwhile Davy had been appointed as assistant lecturer and Gillray's cartoon shows him acting in that capacity to Garnett, with his curly hair and roguish look.

By this time the finances of the Institution were in a bad way, the early enthusiasm was over, subscriptions were falling and expenses rising, but Davy's appointment soon restored its solvency and, while changing somewhat its original purpose, he gave it the character and place it has held in British science for 150 years. Six weeks after his arrival Davy gave the first of a course of lectures on galvanism. It was a brilliant success. The *Philosophical Magazine* reported: 'Mr Davy who appears to be very young, acquitted himself admirably well. From the sparkling intelligence of his eye, his animated manner, and the tout ensemble, we have no doubt of his attaining distinguished excellence.'

The Managers, with a utilitarian outlook, then decided that Davy should give a course on the tanning of leather and he had three months' leave to study the industry. In September he began a series of experiments on the chemistry of the substance employed in tanning, the results of which were published in the *Philosophical Transactions* and gained for Davy a Copley Medal. It cannot be claimed that he made any serious contribution to that complex traditional industry and it was his only venture into organic chemistry.

Davy's real triumph came when he delivered the introductory discourse to a course of lectures on chemistry on 21 January 1802. The lecture theatre was crowded and Davy's moving appeal for interest made a deep impression. He pictured the contribution of chemistry to the arts and sciences, its creative power in the golden age to come and its value to the individual.† Davy's youth, his simplicity, his earnestness and his natural eloquence won universal applause. Some of the founders were not happy at the change which made the Royal Institution a centre of fashion. In 1804 Sir Joseph Banks wrote to Rumford, who was then in Paris, 'It is now entirely in the hands of the profane'.

Maria Edgeworth gives us a shrewd picture of Davy at this time. In a letter dated 8 October 1802, she says: 'The first person we saw was Mr Davy, he is much improved indeed since I saw him last—talks sound sense and has left off being the

---

† Davy had read the proofs of Wordworth's Preface to the first edition of *Lyrical Ballads* with its appeal on behalf of poetry and as Professor Sharrock has pointed out there is a strong resemblance between the aims of Wordsworth and the purpose Davy set himself in his introductory discourse. This in its turn probably influenced Wordsworth's Preface to the second edition of *Lyrical Ballads*.

Humphry Davy

Cosmogony Man. ... After we had seen all the wonders of the Royal Institution, Mr Davy walked with us and got into the depths of metaphysics in the middle of Bond Street, I don't know whether he or the Bond Street loungers amused me most.'

DAVY'S FIRST BAKERIAN LECTURE, NOVEMBER 1806

In his early years at the Royal Institution Davy was not altogether free to develop his own ideas, as a committee decided what researches should be undertaken by the Professor. First there was tanning and then in 1804 the Managers decided to form a collection of minerals and to set up an assay office for the improvement of mineralogy and metallurgy. So Davy had to turn his head to mineral analysis, which was not his natural bent, and he went to Wales and Ireland collecting minerals, doing some fishing at the same time. He had not entirely given up his original plan of becoming a physician as in July 1804, he was admitted a Fellow Commoner of Jesus College, Cambridge, but he soon abandoned the idea as he found it impossible to keep residence. It was only in the intervals of his other duties that Davy was able to continue his researches in voltaic electricity until the autumn of 1806, when he began the experiments which formed his first Bakerian Lecture.

In this he first examines the source of the acids and alkalis which are formed during the passage of a current through water and shows that these are due to the presence of dissolved substances derived either from the containing vessels or impurities. By electrolysing water in gold vessels in a vacuum he showed that only hydrogen and oxygen are formed. He thus removed a certain mystery as to the appearance of acids and alkalis during electrolysis, just as Lavoisier in an early paper had destroyed the myth that water could be converted into earth by showing that it came from the containing vessel. Davy then investigated the transfer of substances which occurs during the passage of a current, and came to the conclusion that the repellent or attractive energies are communicated from one particle to another so as to establish a conducting chain in the fluid.

He then considered the chemical changes produced by electricity and their relation to chemical affinity. Davy's reasoning is quite simple. When bodies are brought into contact and separated they exhibit different electrical states. By means of a voltaic pile it is possible to change the natural electrical state of a substance and by so doing to change its chemical behaviour, e.g. to enhance or eliminate its tendency to oxidation. Hence, argued Davy, may not chemical affinity be due to the opposite electrical states of bodies? 'In the present state of our knowledge, it would be useless to speculate on the remote cause of the electrical energy, or the reason why different bodies, after being brought into contact, should be found differently electrified; its relation to chemical affinity is, however, sufficiently evident. May it not be identical with it, and an essential property of matter?' And then he suggested that the intensity and quantity of the electrical energy capable of destroying equilibrium might be a measure of the chemical affinity—a remarkable prediction.

The publication of Davy's paper created a great sensation. The linking of electricity with chemical affinity brought a new fundamental conception into chemists' minds, the underlying cause of which only became clear a century later. Davy was wise in avoiding further speculation. Berzelius by introducing the idea of polar atoms got himself into serious difficulties.

Napoleon had just founded a medal as a prize to be given for the best experiment each year on the galvanic fluid. The Committee of the Institute awarded the prize to Davy for his Bakerian Lecture. The two nations were then at war. Some thought that Davy should not accept the prize, but said Davy: 'If the two countries or governments are at war, the men of science are not. That would, indeed, be a civil war of the worst description: we should rather, through the instrumentality of men of science, soften the asperities of national hostility.' What would Davy have said to-day?

Davy was elected into the Royal Society in 1803, and in 1807 he became Secretary, an office he held until 1812.

THE DISCOVERY OF POTASSIUM AND SODIUM AND THE ALKALINE EARTH METALS

There is a gap in the laboratory notebook for 1807, but in October Davy was busy with his battery trying to decompose potash and soda. Having failed to do so in solutions, he then tried with solid potash, moistened by contact with the air to make it a conductor, between platinum electrodes and he obtained small globules with a high metallic lustre which burnt in the air. The same result was obtained *in vacuo* and with fused potash, and shortly after a similar substance was obtained from soda. A quick examination of their properties convinced Davy that they were the 'combustible bases of the fixed alkalies' and that in spite of their lightness they were metals. In the laboratory notes he calls them first potagen and sodagen but in the Bakerian Lecture of 19 November 1807 they are called 'potasium' and 'sodium'.

The isolation of these new metals with their remarkable properties and the verification of Lavoisier's prophecy was the most exciting discovery that Davy made. But its great value to him was to provide him with new tools of research, 'powerful agents for analysis' as he called them. Davy was a firm believer in Lavoisier's antiphlogistic theory and now that he had verified Lavoisier's prophecy by isolating the alkali metals, oxygen was not only the principle of acidity but also of alkalescence. Near the end of his Bakerian Lecture of 1807 he is looking for the oxygen that is responsible for the alkalinity of ammonia and some faulty experiments that are strangely reminiscent of his 'infant chemical speculations' convince him of its presence. So that, like his Swedish contemporary Berzelius, he was a victim of Lavoisier's over-generalization. But Davy also had running through his mind the idea that perhaps there was something after all in the old phlogistic theory. Might not all metals contain hydrogen or some common principle? His electrochemical work seemed to give this some support.

'Oxygen is the only body which can be supposed to be elementary, attracted by the positive surface in the electrical circuit, and all compound bodies, the nature of which is known, that are attracted by this surface, contain a considerable proportion of oxygen. Hydrogen is the only matter attracted by the negative surface which can be considered as acting the opposite part to oxygen; may not the different inflammable bodies, supposed to be simple, contain this as a common element?'

Davy was not alone in these doubts and speculations, which he tried to solve by experiment. Berzelius in Sweden and Gay-Lussac and Thenard in France were hard at work too.

Almost immediately after the Bakerian Lecture on potassium and sodium Davy was seriously ill and his health was a matter of public anxiety. By April 1808 he was back in the laboratory once again trying to extract oxygen from muriatic acid and next he was trying to decompose the alkaline earths with the battery, with small success until a letter came from Berzelius telling him that he had succeeded using a mercury electrode. By distilling the amalgams Davy quickly isolated two metals which he named barium and strontium. But Berzelius's letter also contained the news of his discovery of ammonium amalgam, that puzzling substance which added to Davy's perplexity. In his reply to Berzelius he says (July 1808): 'Your discovery of the amalgamation of the basis of ammonia has afforded me the highest of pleasure. The fact is no less new and unexpected, than extraordinary and important. . . . May not hydrogene and nitrogene be metals in the state of elastic vapour? Should this not be the case and your brilliant hypothesis of the composition of metals be true, we may hope at some period for a rational Alchemy.'

Meanwhile Gay-Lussac and Thenard had obtained potassium in much larger quantities by heating potash with iron and were also investigating its properties. By heating it with ammonia they had obtained hydrogen and potassamide, which on treatment with water gave ammonia and potash. This they thought proved that Davy's potassium was not an element but a compound of potash and hydrgoen. Davy quickly replied with experiments to show that this view was untenable and eventually Gay-Lussac and Thenard agreed with him.

DOUBTS ABOUT THE ELEMENTS

For two years after after the isolation of the alkaline earth metals Davy was examining 'the action of potassium and sodium on bodies not hitherto decompounded with very extraordinary results', as he wrote to Berzelius in November 1808: 'These experiments seem to prove the existence of oxygene in muriatic, boracic and fluoric acids and render it probable in hydrogene, nitrogene, the diamond, sulphur and phosphorus.' It is not unfair to say that at this time Davy's judgement was warped by his adherence to the oxygen theory of acids and alkalis; some of his experiments were faulty and his interpreta-

Humphry Davy

tions of the results is reminiscent of the phosoxygen days. In March 1809 he writes to Berzelius; 'From nitrogen I have obtained a large quantity of oxygene; but as yet I have not been able to conclude concerning its basis. It seems resolved into nothing but oxygene and hydrogene; either some minute quantity of matter has escaped my research, or water and ammonia consist of the same kind of elementary matter; and consequently oxygene, hydrogene and nitrogene may be different modifications of the same substance under different electrical states, or in different states of combination with imponderable matter.'

During these years there was keen rivalry between Davy and Gay-Lussac and Thenard who were working in the same field in Paris with a large voltaic pile and with large quantities of potassium made by the reduction of potash by iron. There were inevitably questions of priority as, for instance, the isolation of boron, and often opposing views. Davy's experiments at this period give the impression of being hurried and they lacked the disciplined quantitative methods of the French chemists. Many discoveries were made by both and undoubtedly Davy, in spite of his rather wild speculations, was getting first-hand knowledge of many substances for the first time, and gaining that intuitive sense in which he excelled of the similarities between the members of groups of elements.

His Bakerian Lecture of 1809 describes the discovery of telluretted hydrogen, unsuccessful attempts to decompose silica, alumina, and beryllia, and inconclusive experiments with nitrogen, ammonia, and ammonium amalgam. The search for oxygen and hydrogen in other elements was the will-o'-the-wisp Davy was chasing so eagerly. The run of his mind is shown by an entry in the laboratory notebook on 26 August 1809. 'Should Tellurium *really* possess a strong attraction for hydrogen it will be the most valuable discovery of a reagent ever made. It will lead to test facts of the Phlogistic and Antiphlogistic theories.'

Davy's perplexities arose from two ideas that were running through his mind, the first that all metals and inflammable bodies contained hydrogen, the second the generally accepted doctrine that oxygen is contained in all acids to which he himself had added evidence that it was contained in all alkalis. It was the rigid acceptance of this doctrine that was responsible for so much make-believe in chemists' minds.

CHLORINE

Davy once spoke of himself as not too much 'devoted to consistency'. It seems suddenly to have dawned on him that all the experimental evidence indicated the absence of oxygen in the substance Lavoisier had called oxymuriatic acid and that the only reason for assuming its presence was analogy. In a paper read before the Royal Society in July 1810 he described most convincingly the experiments which had failed to reveal the presence of oxygen. Charcoal at a white heat had no action on oxymuriatic acid. High-tension sparks left it unchanged. Its compound with tin on treatment with ammonia volatilized leaving no tin oxide. If it contained oxygen, its compounds with phosphorus should contain muriatic acid and phosphoric acid, and on treatment with ammonia yield the corresponding salts. But the experiment yielded a non-volatile solid, phospham, of quite different properties.

Davy then points out the weakness of the evidence usually given to prove both the presence of oxygen in oxymuriatic acid and of water in muriatic acid. Oxygen is only obtained from oxymuriatic acid in presence of water and as it combines with hydrogen to form muriatic acid the oxygen is naturally displaced. Gay-Lussac and Thenard had estimated the water in muriatic acid by passing it over lead oxide to form lead muriate and measuring the water produced. Davy points out that this is formed from the oxygen of the oxide and the hydrogen of the muriatic acid. Metals if heated in muriatic acid gas give a muriate and half the volume of hydrogen, the same volume as is found by their action on water. 'Few substances', Davy sums up, 'perhaps have less claim to be considered as acid, than oxymuriatic acid. As yet we have no claim to say that it has been decompounded; and as its tendency of combination is with pure inflammable matters, it may possibly belong to the same class of bodies as oxygen.'

Gay-Lussac and Thenard, it is true, had considered this possibility a year earlier but had rejected it in favour of the older view based on analogy.

Probably Davy's greatest service to chemistry was his recog

Humphry Davy

nition of the elementary nature of oxymuriatic acid to which he gave the name chlorine. He thus broke with a doctrine generally accepted by chemists, which in its way was almost as misleading as phlogiston. And the recognition of chlorine as an element, or as Davy put it 'undecompounded', carried with it the recognition of muriatic acid as a compound of hydrogen and chlorine, the germ of an entirely new conception, the idea of acids as compounds containing hydrogen which could be replaced by metals to form salts.

Davy alone had the courage and the realism to break with a tradition backed by the great authority of Lavoisier. His objective view cleared many doubts from chemists' minds and was the first step towards sweeping away the artificial idea of the dualistic nature of salts. As evidence accumulated, chemists sided with him but it was ten years before Berzelius was converted to chlorine and even then dualism remained a millstone round his neck.

DAVY'S LECTURES

Much of Davy's time at the Royal Institution was given to the preparation of his lectures. He took immense pains with them, each lecture being written afresh usually the day before it was delivered. The manuscripts in his clear flowing hand show the ease of his composition and are a curious contrast to the hasty untidy notes which record his impulsive experiments. It was his usual custom to rehearse each lecture with his assistants to ensure that the experiments ran smoothly and to practise the emphasis and intonation of his delivery. His brother says of him: 'His manner was perfectly natural, animated and energetic, but not in the least theatrical. In speaking he never seemed to consider himself as an object of attention; he spoke as if devoted to his subject, and as if his audience were equally devoted to it and their interest concentrated on it. The impressiveness of his oratory was one of its great charms.'

Davy was recognized as the greatest living exponent of chemistry and for twelve years the lecture theatre of the Royal Institution was filled with crowded audiences eager to have the latest news of Davy's own discoveries.

And Davy's lectures were not limited to chemistry; geology

was one of his favourite subjects. Dr. Paris describes the final scene in a lecture on volcanoes at a time when Davy thought that the alkaline earth metals might be responsible for volcanic action. 'A mountain had been modelled in clay, and a quantity of the metallic bases introduced into its interior: on water being poured upon it, the metals were soon thrown into violent action —red hot lava was seen flowing down its sides, from a crater in miniature—mimic lightning played around, and in the instant of dramatic illusion, the tumultuous applause and continued cheering of the audience might almost have been regarded as the shouts of the alarmed fugitives of Herculaneum or Pompeii.'

For ten years Davy gave at the invitation of the Board of Agriculture an annual course on agricultural chemistry, and his lectures were published in 1813. He had made a careful study of the problems of agriculture and his frequent journeys had brought him into touch with agricultural practice in many countries. In the lectures on plant growth and plant metabolism, soils and fertilizers, he showed very clearly the help that chemistry could give, and their great virtue was to insist that agriculture must look to science for the explanation of its problems. Davy was particularly interested in nutrition and in the nutritive value of different crops and he encouraged valuable systematic work on grasses at Woburn. He often referred to the experiments he was conducting, for which he was given facilities by several of his friends, but his own work was rather hasty and superficial and it cannot be said that he made any outstanding contributions to agricultural science. He, like most of his contemporaries, thought that plant growth depended on the absorption by the roots of organic substances in solution in the soil, rather than on the absorption of carbon dioxide in the air and his studies of soils and fertilizers failed to recognize the importance of nitrogen, phosphate, and potash. But, as always, he made many shrewd comments and suggestions, particularly in soil physics.

Undoubtedly his lectures and his frequent visits to the farms of landowners like Mr Coke of Holkham and the Duke of Bedford did much to stimulate a more scientific approach to farming problems and to initiate controlled experiments. Davy made scientific farming fashionable.

Humphry Davy

'ELEMENTS OF CHEMICAL PHILOSOPHY', 1812

Fortunately we have a record of Davy's views of chemistry as a whole in his *Elements of Chemical Philsophy*, which was published in 1812. As Berzelius said, it showed that Davy had studied chemistry more by experiment than by reading as it is based so largely on his personal experience. In the general laws of chemical combination he accepts the laws of combination in definite and multiple proportions without committing himself to the atomic theory. He uses the word proportion instead of the atomic weight and he assumes that the volume ratios in which elements combine indicate the number of proportions of each. Thus in water he assumes two proportions of hydrogen to one of oxygen, so that the numbers representing the proportions are in the ratio of their specific gravities, hydrogen 1, oxygen 15. The proportional number of potassium he takes as about 75.

In the chapter dealing with electrical attraction and repulsion and their relation to chemical changes he explains that the possibility of the dependence of electrical and chemical action upon the same cause has been much misrepresented. His hypothesis was not that chemical changes were occasioned by electrical changes, which are 'considered, on the contrary, to be distinct phenomena; but produced by the same power, acting in one case on masses, in the other case on particles'.

He classes oxygen and chlorine together as empyreal undecompounded substances, and hydrogen, azote, sulphur, phosphorus, carbon, and boron as undecompounded inflammable or acidiferous substances. Of all gaseous substances, hydrogen, he says, is most distinctly characterized as an element. He accepts azote as undecompounded but he still regards its chemical nature as unknown. The metals represent a chain of bodies with gradations of resemblance. 'We know nothing of the true elements belonging to nature' and he hints at the idea that all inflammable matters may be found to contain hydrogen. The existence of ammonium amalgam gives support to this view, and the amount of the hydrogen in metals may be sufficient to account for their combinations with oxygen and chlorine. This would be in harmony with the theory of definite proportions.

He ends with the suggestion that: 'A few undecompounded

bodies, which may, perhaps, ultimately be resolved into still fewer elements, or which may be different forms of the same material, constitute the whole of our tangible universe of things. By experiment they are discovered even in the most complicated arrangements, and experiment is, as it were, the chain that binds down the Proteus of nature and obliges it to confess its real form and its divine origin.'

Berzelius was in England when the book appeared and Davy sent a copy saying: 'I shall be glad to profit by your critical observations in a second edition for which I am now preparing.' Berzelius, with his encyclopaedic knowledge and perhaps a little nettled by Davy's cavalier treatment of him, took the invitation seriously. He sent Davy notes on over a hundred passages which contained errors of fact or in which he differed from Davy's interpretation. Unfortunately he criticized sarcastically a quantitative determination over which Davy had taken unusual pains and which was, in fact, particularly accurate. Of course the main quarrel between them was over chlorine. Davy in his reply said: 'I once had an hypothesis that hydrogene, oxygen and azote were different forms of water, you justly objected to this hypothesis. I object to your hypothesis: oxygen plays in your system the part that hydrogene played in the phlogistic system. Oxygene with me is not the exclusive neutralizing or negative principle.'

Davy was mortally offended and their correspondence was broken off for eight years, just when Davy's realistic judgement, which was so often right, might have been such a help to Berzelius.

MARRIAGE AND THE GRAND TOUR

By 1812, at the age of thirty-three, Davy was the most celebrated chemist in Europe and in this year great changes came into his life. In the second week of April he was knighted by the Prince Regent at a levee, he gave the final lecture of his last course at the Royal Institution and two days later he married Mrs Apreece, a rich widow and a distant cousin of Sir Walter Scott. Scott describes her in his *Journal* as: 'gay, clever and most actively ambitious to play a distinguished part in London society. . . . As a lion catcher I could pit her against the world.'

Humphry Davy

In May Davy told the Managers that he could not pledge himself to deliver lectures but he was willing to accept the office of Professor of Chemistry and Director of the Laboratory without salary, and so his connection with the Royal Institution continued. From the day of his marriage Davy's time was much occupied by rounds of country visits and journeys abroad and he seldom had long continuous spells of work in the laboratory. But his papers though few in number are significant both in their subjects and in showing the power of his mature mind.

In June 1812 Davy read a paper 'On some combinations of phosphorus and sulphur' in which he describes the two chlorides of phosphorus which he had discovered in 1809. He showed that the lower chloride reacts with water to give phosphorous acid, from which on heating he obtained a new hydride of phosphorus. He is now interested in the 'laws of definite combination' which he illustrated by some analyses. His figures for phosphorous trichloride are accurate, but he was wrong in finding that the higher chloride contained twice as much chlorine as the lower, an error which he transferred to phosphoric and phosphorous acids and repeated again in 1818.

In November he had a short paper 'On a new detonating compound', nitrogen chloride, which had just been discovered in Paris, warning chemists of its explosive character as he had been severely wounded by fragments of glass entering his eye from the explosion of a quantity 'scarcely as large as a grain of mustard seed'.

In 1813 Davy, 'struck by the analogy between the oxymuriatic and the fluoric compounds', began a fresh investigation of the latter. Meanwhile Ampère, an early convert to his chlorine theory, wrote to him suggesting that the fluoric compounds contained what Davy called 'a peculiar undecompounded principle, analogous to chlorine and oxygen' for which Ampère suggested the name fluorine. On the old theory fluoric acid like sulphuric and nitric acids contains water, on the new it is a compound of hydrogen and fluorine. Davy prepared pure fluoric acid in platinum vessels and failed in various experiments, similar to those he made with muriatic acid, to detect either water or oxygen in it. He tried by various means to decompose it into hydrogen and fluorine, but failed to isolate the latter owing to its affinity for other elements. Finally, he decided in favour of Ampère's view and in a second paper he carries out quantitative experiments to determine the number representing the definite proportion in which fluorine combines. By analysing calcium fluate and potassium fluate he arrives at a number, about thirty-three, for fluorine on the supposition that liquid fluoric acid consists of two proportions of hydrogen to one of fluorine. He then shows that this is consistent with the composition of other fluorine compounds. He evidently attaches weight to this, but the word atom is carefully avoided.

It is characteristic of his tentative approach that he sums up: 'In the views that I have ventured to develop, neither oxygen, chlorine nor fluorine, are asserted to be elements; it is only asserted, that, as yet, they have not been decomposed.'

He ends with the conjecture that there may be other more subtile bodies of this class, and that the presence of an extremely light and subtile principle, which has hitherto escaped detection may account for the differences between the diamond and charcoal. Some chemical difference must exist between them.

Davy was essentially a soloist. He had had no pupils and no co-authors, but early in 1813 his historic partnership began. A young journeyman bookbinder, Michael Faraday, was taken by a member, Mr Dance, to hear some of Davy's last lectures at the Royal Institution. He made a fair copy of the notes he had taken, bound them in leather and sent them to Davy saying that he wished for scientific work. Davy sent an encouraging reply and two months later by a lucky stroke of providence Mr Payne, the laboratory assistant, having quarrelled with Mr Newman the instrument maker assaulted him, and Faraday thanks to Davy, was appointed in his place. From then onwards Faraday helped Davy in most of his experiments and often made the fair copies of papers for publication. It was a great experience for Faraday for which he was always grateful.

When Davy and his wife left England for their grand tour of Europe in October 1813, Faraday went with them in the awkward position of secretary-valet. It is clear from his diary that even then the marriage was not going too smoothly. But Faraday put up with the difficulties of his position. 'The constant presence of Sir Humphry Davy', he wrote to a friend 'is a mine inexhaustible of knowledge and experience.' The went first to Paris, Davy having a safe-conduct although the two countries were at war. There Ampère gave him some of the new substance which Courtois had found in kelp. Davy examined its properties with the help of a small case of chemical apparatus he carried with him and from the close resemblance of its compounds to those of chlorine and fluorine, he recognized it as a member of the same group and named it iodine from the colour of its vapour. The accuracy and completeness of his qualitative study made in a short time with very simple apparatus shows Davy's experimental skill, his keen eye and his grasp of essentials in chemical behaviour. He isolated many of the common compounds of iodine with metals and non-metals, discovered the iodates and studied the reactions of hydriodic and iodic acids. Gay-Lussac was hard on the same scent and the old rivalry was aroused. Davy, thinking that Gay-Lussac was claiming some of his ideas, sent a note to Cuvier and on 10 December a paper to Banks asking for it to be published as quickly as possible. 'The subject has occupied me constantly during the last three weeks.' He added in a postscript: 'M. Cuvier in the report of the labours of the French Inst. has mentioned that I communicated a paper to the Institute. I suppose he must mean the letter I wrote to him and which he read and which was published in the *Journal de Physique*. It was my intention not to have published a word in French on the subject, but a sense of justice obliged me to do so. I found my opinions in one instance so freely made use of without acknowledgement that, to prevent the facts given in my paper to the Society from being anticipated in the *Moniteur*, I was obliged to say in a few words what I had done.' Later from Florence and from Rome he sent home notes on iodine pentoxide which he had obtained pure by the action of euchlorine on iodine. He criticized Gay-Lussac for calling the pentoxide iodic acid. 'I am desirous of marking the *acid* character of oxyiod combined with water. . . . It is not at all improbable that the action of hydrogen in the combined water is connected with the acid character of the compound.'

He returns to the same point in a short note in 1816 on Analogies between the undecompounded substances and the nature of acids'. In this he attacks Gay-Lussac, who has now accepted his view of chlorine and iodine but classes them with sulphur rather than oxygen. Davy objects to this and to Gay-Lussac's view of hydrogen as an alkalizing principle and of azote as an acidifying principle. It is strange, he says, that hydrogen forms some of the strongest acids and azote is nine-tenths of the weight of the volatile alkali. 'This is an attempt to introduce into chemistry a doctrine of occult qualities, and to refer to some mysterious and inexplicable energy what must depend on a peculiar corpuscular arrangement.... When certain properties are found belonging to a compound, we have no right to attribute these properties to any of its elements to the exclusion of the rest, but they must be regarded as the result of combination.'

Davy then attacks the oxide theory of acids and dualism. 'An acid compound of five proportions of oxygen and one of nitrogen is altogether hypothetical, and it is a simple statement of facts to say that liquid nitric acid is a compound of two proportions of hydrogen, one of azote and six of oxygen ... There are very few of the substances which have been always considered as neutral salts, that really contain the acids and alkalies from which they have been formed. The muriates and fluates must be admitted to contain neither acids nor alkaline bases.... the substitution of analogy for fact is the bane of chemical philosophy; the legitimate use of analogy is to connect facts and to guide to new experiments.'

This is Davy's final comment on the use of analogy which had led him astray for so many years.

He came tantalizingly near to the unitary theory of compounds and the hydrogen theory of acids which had to wait many years before they were generally adopted. As was his habit he left them as tentative suggestions. His lack of formal training, his many interests, and his temperament stood in the way of his taking a systematic view of chemistry as a whole and carrying his suggestions to their logical conclusions.

Davy wrote two other papers while he was abroad. In Florence and in Rome he burnt diamonds with a burning lens in globes of oxygen and found the volume of carbon dioxide produced was equal to that of the oxygen consumed. This satisfied him that the diamond contained no other element than carbon and differed from charcoal only in its crystallization, as Tennant had said. In Rome, with the help of Canova

Humphry Davy

who was in charge of ancient works of art, he made an extensive examination of the pigments used in classical mural paintings and in vases. Davy had evidently kept his knowledge of the classics in good repair as he quoted Vitruvius, Pliny, and Theophrastus, and he was always fond of a classical quotation. The paper shows his keen interest in classical art.

THE SAFETY LAMP

Davy returned from his long foreign tour in May 1815 and we come now to the invention of the safety lamp for which his name is so widely known.

By 1813 the increasing number of colliery explosions as the mines went deeper became a great source of anxiety and, after an explosion in which ninety-two lives were lost, a society was started to find means of combating the danger. Davy was already on the continent, but on his return in 1815 his help was enlisted as no solution had been found. In August he went to Northumberland and studied the problem. After a round of visits in September he returned to London and early in October began experiments at the Royal Institution with some specimens of fire damp sent from Newcastle. He first analysed them and found that they consisted of methane mixed with varying amounts of air. He next determined the amounts of air with which methane must be mixed in order to form an explosive mixture. He then compared the inflammability of these mixtures with those of other inflammable gases. The next step was 'to examine the degree of expansion of mixtures of fire damp and air during their explosion and likewise their power of communicating flame through apertures to other explosive mixtures'. Davy found that the flame of an explosive mixture of coal gas and air moved slowly down a tube $\frac{1}{4}$ inch in diameter and would not ignite in a tube $\frac{1}{7}$ inch in diameter. Experiments with fire damp and air gave similar results, small metal tubes being more effective in stopping the explosion than glass. Metal tubes $\frac{1}{5}$ inch in diameter stopped the explosion if they were $1\frac{1}{2}$ inch long. This, Davy saw, was due to the heat lost in contact with the cooling surface, which brings the temperature below that needed to fire the other portions. He found too, that explosions would not pass through very fine wire gauze and that a small addition of nitrogen or carbon dioxide to an explosive mixture deprived it of the power of explosion. 'The consideration of these various facts', said Davy, 'led me to adopt a form of lamp in which the flame by being supplied with only a limited quantity of air should produce such a quantity of azote and carbonic acid as to prevent the explosion of fire damp, and which by the nature of its apertures for giving admittance and exit to the air, should be rendered incapable of communicating any explosion to the sounding air.'

These experiments show Davy at his best as a brilliant investigator, in the directness of his attack, in the logical sequence of the experiments, and their swift result. In a fortnight he had arrived at the principle of the safety lamp which was to mean so much to the future expansion of coal mining. On 30 October he sent a description of three forms of safety lamp to Newcastle saying: 'My results have been successful far beyond my expectations. . . . I wish you to examine the lamps I have had constructed before you give any account of my labours to the Committee.' On 9 November he read his first paper on the subject to the Royal Society 'On the fire-damp of coal mines, and on the methods of lighting the mines, so as to prevent its explosion.'

By this time oil lamps had been made to his design with an air-tight glass round the flame to which air was admitted from below through narrow metal tubes or through concentric metal cylinders, the exit at the top of the glass being through similar cylinders or metal gauze. He had satisfied himself by a series of experiments in explosive mixtures that the lamp would continue to burn without transmitting the explosion.

On 18 January he read two more papers to the Royal Society describing the final form of the safety lamp in which the glass is replaced by a wire gauze cylinder which is attached by an air-tight joint to the oil container of the lamp. With iron wire gauze of 576 apertures/in^2 the lamp burnt safely in explosive mixtures even when the gauze became red-hot, and Davy reported that the cylinder lamps had been tried in two of the most dangerous mines near Newcastle with perfect success.

Mr Buddle wrote after the test carried out at the Hebburn Colliery on 17 January 1816:

To my astonishment and delight, it is impossible for me to express

Humphry Davy

my feelings at the time when I first suspended the lamp in the mine and saw it red hot; if it had been a monster destroyed I could not have felt more exultation than I did. I said to those around me 'We have at last subdued this monster'.

But Davy's was not the only safety lamp. George Stephenson, then an engine tenter at the Killingworth Colliery, had invented a lamp consisting of an oil burner with a glass chimney, the air entering and leaving by small apertures in metal disks. Stephenson by native wit and empiricism had thus made a lamp similar to Davy's original pattern, except that it lacked the exact knowledge of the behaviour of explosive mixtures which was the basis of Davy's design. The first trial of Stephenson's lamp had taken place on 21 October 1815, and there were soon two parties advocating the rival claims of the two inventions. The weakness of Davy's lamp in its original form was that it was unsafe in an explosive atmosphere blowing past it at a rate exceeding 6–7 ft/s. Davy had warned users of this and had added the protection of a metal shield. Stephenson's lamp also was unsafe in air currents exceeding 8 ft/s as a down draught was set up and the flame passed to the surrounding atmosphere. As Davy pointed out, the apertures at the bottom of Stephenson's lamp were four times too large and those above twenty times too large for safety. Later Stephenson adopted Davy's gauze as additional protection, Davy having refused to patent his invention when urged to do so. 'My sole object', he wrote to Mr Buddle, 'was to serve the cause of humanity; and if I have succeeded I am amply rewarded in the gratifying reflection of having done so. It might undoubtedly enable me to put four hourses to my carriage; but what would it avail me to have it said that Sir Humphry drives his carriage and four.'

The use of Stephenson's 'Geordie' was confined to a few pits in Northumberland, while Davy's lamp, with progressive improvements, was ultimately used wherever dangerous conditions were encountered. In October 1817 Davy was presented with a service of plate by the coal owners of Northumberland as a token of their gratitude for the invention which he had so unselfishly placed at their disposal. By Davy's will this service of plate passed on Lady Davy's death to his brother John, and in the event of his having no heirs in a position to make use of it, it was to be melted and given to the Royal Society 'to found a medal to be given annually for the most important discovery in chemistry anywhere made in Europe or Anglo-America'. So the Davy Medal, which has been awarded since 1877, owes its origin to the mine owners' gratitude for the safety lamp.

But public controversy over the claims of the two lamps still continued and it was hinted that Davy knew of some unpublished work of Smithson Tennant, who had discovered that explosions of mixtures of coal gas and air would not pass through narrow tubes. In November Sir Joseph Banks, P.R.S., Brande, Hatchett, and Wollaston issued a statement that: 'Sir Humphry Davy not only discovered independently of all others, and without any knowledge of the unpublished experiments of the late Mr Tennant on flame, the principle of the non-communication of explosions through small apertures, but that he also has the sole merit of having first applied it to the very important purpose of a safety lamp.'

Twenty years later the Report of the Select Committee appointed by Parliament in 1835 'to enquire into the nature, cause and extent of those lamentable catastrophes that have occurred in the Mines of Great Britain' bears eloquent testimony to the contribution of the safety lamp to mining development.

One striking fact requires to be particularly pointed out. If the year 1816 is assumed as the period when Sir Humphry Davy's lamp came in to use, a term of 18 years since 1816, and a similar term prior to 1816 being taken it will be seen that in the 18 years previous to the introduction of the lamp 447 persons lost their lives in the counties of Durham and Northumberland whilst in the latter term of 18 years the fatal accidents amounted to 538. To account for this increase it may be sufficient to observe that the quantity of coal raised in the said counties has greatly increased; seams of coal, so fiery as to have lain unwrought have been approached and worked by the aid of the safety lamp. Many dangerous mines were successfully carried on, though in a most inflammable state, and without injury to the general health of the people employed in them. Add to this the idea entertained that on the introduction of that lamp the necessity for former precautions and vigilance in great measure ceased.

Your Committee have endeavoured to investigate with strict

Humphry Davy

impartiality the merits of the different lamps which have been brought under their notice. In the course of the Evidence many varieties will be found described. The invention claimed by the late Sir Humphry Davy, on principles demonstrated by that able philosopher, may be considered as having essentially served the mining interests of this kingdom, and through them contributed largely to the sources of national as well as individual wealth. Many invaluable seams of coal never could have been worked without the aid of such an instrument; and its long use throughout an extensive district, with the comparatively limited number of accidents, proves its claim to be considered, under ordinary circumstances, a Safety-lamp.

It was Davy's great service to have established by systematic experiments the principle of the protection given by wire gauze of suitable dimensions. That has remained the basic feature of oil burning safety lamps from 1816 until today, when in spite of the conveniences of the electric safety lamp 20 per cent of the miners' lamps in use in this country are still of the Davy type. They not only give light but they are the only practical means of detecting immediately the presence of fire damp, of estimating within certain limits its concentration, and of indicating when the oxygen in the air falls below 19 per cent. They are widely used also for similar purposes in the United States.

Subsequently Davy continued his researches into the nature of flame, which he regarded as the combustion of an explosive mixture of gas and air. The luminosity of a coal-gas flame he attributed to the presence of particles of carbon. He examined the diminution of pressure at which various flames are extinguished, which he explains by the lowering of the temperature, since if heated they will burn at lower pressures. He then attempted to compare the heats of combustion of various gases by burning small flames under a small copper calorimeter containing oil. In studying the explosion of mixtures of hydrogen and oxygen, he found that under certain conditions of temperature and pressure they combined slowly. He then examined the amounts of various gases required to prevent the inflammation of explosive mixtures of hydrogen and oxygen, which he attributed to the effect of cooling, just like the wire gauze in the safety lamp. Finally, he discovered the power of heated platinum to bring about the slow combination of mixtures of vapours and air below their ignition temperature. Faraday was helping him with these experiments and years later he returned to the investigation of the catalytic properties of platinum.

PRESIDENT OF THE ROYAL SOCIETY

In May 1818 Davy set out on another continental tour, one object being to find means of unrolling papyri at Herculaneum; in this he was fairly successful but the results proved to be of little value. He found Vesuvius in action and he paid numerous visits to the crater, making a careful investigation of the gases and solids ejected during the eruption. The results were published in the *Philosophical Transactions* in 1828.

On the return journey, hearing in Paris that Banks, for forty-two years President of the Royal Society, was so ill that he could not long survive, Davy set out at once for London where he arrived on 6 June 1820. Banks died on Monday, 19 June, and active discussions started at once among the Fellows as to the choice of his successor. Davy's high-handed manner had antagonized many of them who would have preferred Wollaston, if he could be persuaded to stand. Other candidates were mentioned including Prince Leopold, but the real issue lay between Davy and Wollaston. A memorandum written at the time by Sir John Herschel tells the story of the eventful week after Banks's death.

On Tuesday Herschel and Babbage, after consulting other Fellows, called on Wollaston who said he had decided not to stand. Nevertheless, if convinced it was for the good of the Society, he could hardly say what he might not do. Wednesday was spent in getting further support for Wollaston in order to put pressure on him, but on Thursday he was still undecided. In the evening, Davy, Wollaston, and a number of the Fellows met at the Club Dinner at the Crown and Anchor. When Davy was told that Wollaston had many supporters he asked him for a private talk, when he taxed him with canvassing. Wollaston then determined to oppose him, but later in the evening he tried to avoid a contest by suggesting the spin of a coin. Davy refused, saying he thought his chances too good to be hazarded. Then they agreed on arbitration, but finally Wollaston withdrew.

Wollaston acted as President until 30 November when Davy was elected, only a few votes being cast for the other candidate, Lord Colchester.

Davy's heart was set on the Presidency as he saw in it a great opportunity to advance the cause of science and to promote the interests of the Society. Davy's dream was to see the Society controlling the Observatory at Greenwich, a new British Museum for Natural History, and national laboratories amply equipped for original inquiry. It was a great disappointment to him that he could get no effective support from the Government for these projects in spite of the sympathetic attitude of Robert Peel, the Principal Secretary of State in Lord Liverpool's cabinet.

In June 1823 Peel writes to Davy expressing 'sanguine hope that when the rooms at present occupied by the Lottery Establishment shall have been vacated additional accommodation will be provided for the Royal Society'. However, in November 1825 the Lottery Office is still in possession and space is urgently needed for a joint investigation by the Royal Society and Board of Longitude 'on the perfection of glass for nautical and astronomical purposes to endeavour to recover a brand of manufacture which had left the country for Germany and Switzerland'.

Peel shares Davy's views about the British Museum. In December 1824 he writes: 'I own that what with Marbles—butterflies—statues—Manuscripts—Books and pictures I find the Museum is a farrago that distracts attention. There is no division of labour there,' Davy replies: 'Could the British Museum be divided into three distinct departments each with a separate government every thing would become easy. A great Public Library—A Gallery of Art—A Gallery of Science. We have had no great naturalist since Ray which I believe is a good deal owing to this that Britain affords no means of studying Natural History.' But Davy got no support and the opening of the Natural History Museum had to wait for sixty years. In 1826 he wrote: 'I have been twice at the British Museum, but I despair of anything being done there for natural history. The Trustees think of nothing but the arts, and money is only obtained for these objects.'

However, he was successful in other directions. The idea of a zoological collection came from Stamford Raffles, who discussed it with Banks in 1817. In July 1824 there was a meeting in London of friends of 'a proposed Zoological Society' who appointed a Committee, including Davy, with Raffles as Chairman. Davy then sought Peel's support, who wrote him a private letter in December warmly supporting the scheme and asking for details before he raised it with the Prime Minister.

Considering the riches of our country—its vast Colonial Possessions, including almost every variety of climate and every species of natural production—our means and our habits of exploring those parts of the globe which offer no temptation to fixed settlement—but still abound with much that is curious and valuable to the lovers of Natural History—we ought to be ashamed of the beggarly account of Boxes—almost worse than empty, which comprize our specimens of animal life. . . . I should feel proud of contributing my humble efforts to rescue this country from what I think is a just imputation of indifference and neglect.

Davy, who wrote the original prospectus of the Society, sent Peel a long reply setting out his ideas which were mainly directed to the acclimatization of animals, birds, and fishes living in the same latitudes as the British Isles. 'Eight or ten races of partridges, half as many of pheasants.' He suggests 100 to 150 acres near London are needed. 'My feeling is that as much as is possible should be done by contributions but *encouragement* from the Government would certainly be very desirable such as the grant of a piece of ground.' It must be entirely separate from the British Museum.

Raffles in 1825 writing about the scheme for a Grand Zoological Collection says: 'Sir Humphry Davy and myself are the projectors, and while he looks more to the practical and immediate utility to the country gentleman, my attention is more directed to the scientific department.'

Davy's influence with Peel was no doubt a great help towards getting the grant of land in Regent's Park for the Zoological Society in April 1826.

In 1824 Davy was the co-founder with Croker of the Athenaeum.

Geographical exploration was another of Davy's projects. In July 1826 Maria Edgeworth wrote to a friend, 'Yesterday when I am down to breakfast, I found Sir Humphry with a coun-

Humphry Davy

tenance radiant with pleasure and eager to tell me that Captain Parry is to be sent on a new Polar expedition.'

Davy's creative mind was always at work seeking new ways of adding to the dignity and usefulness of the Royal Society. His great ambition was to bring it into more intimate relations with the State. In 1825 the Royal Medals were founded as a mark of Royal favour for the Society.

Marcet in a letter to Berzelius in 1822 says,

> Our friend Davy is acquitting himself well in the presidency; I have noticed a great change in him for the better, thanks to his goodness of heart and to the frankness of his friends, who have let him see that he gains nothing by being high-handed. He presides well at the meetings although a little gauche and too full of ceremony. He comes always in court dress, with a lace jabot and a three cornered hat, the point of which looks up to heaven like the parish beadle's, showing that the wearer was never a soldier or a courtier. But he is full of zeal and devotion to the business of the Society, and he manages still to work from time to time in the laboratory of the Royal Institution.

Davy's first paper after his election as President dealt with his work on the papyri at Herculaneum, and another with his examination of the water found occasionally in cavities in crystals of quartz and other minerals. In 1820, with his usual opportunism, he had followed up quickly the announcement of Oersted's discovery of the magnetic properties of a conductor carrying a current and in a letter to Wollaston he described a number of experiments made with his usual ingenuity. In one of them he showed that a steel needle could be magnetized by the discharge from a Leyden battery, if placed transversely to the wire connected to the battery. He speculated about the earth's magnetism being caused by electrical currents due either to its motions, to internal chemical action, or its relation to solar heat. Strong electrical currents following the apparent course of the sun would account for it. Do not the auroras as the poles depend on electricity?

This he followed up in 1821 by showing the effect of a magnet on the carbon arc and on discharges *in vacuo*. He then did some interesting experiments by comparing the electrical conductivity of various metals and examining the effects of temperature, the cross-section and the length of the conductor. His only means of measuring the strength of the current was the time taken to discharge a battery. He distinguished between the quantity and the intensity of the electricity, and if only he had had some means of measuring the momentary current strength, his clear grasp of the problem might have led to important advances.

His next investigation was an attempt to throw further light on the nature of electricity and its relation to magnetism, light and heat by examining the electrical discharge in a vacuum. By boiling mercury repeatedly in one limb of a U-tube with a sealed-in platinum electrode, he obtained a fairly perfect vacuum and observed the change in the discharge through mercury vapour over a wide temperature interval, and also the effect of admitting small quantities of air. He concluded that the light generated in electrical discharges depends *principally* on some properties of the ponderable matter through which it passes and that space, when there is no appreciable quantity of matter, is capable of exhibiting electrical phenomena.

In 1823 he suggested that Faraday should heat some crystals of chlorine hydrate in a sealed tube. This resulted in the liquefaction of chlorine and was followed by the liquefaction of a number of other vapours. Davy wrote a short note at the time saying that one of his principal objects in these experiments was to see if the changes with temperature of the vapour pressure of some of these substances might not enable them to be used as substitutes for steam in an engine. Davy's idea was that owing to the rapid rise with temperature of the vapour pressure of a substance like carbon dioxide, 'the mere difference in temperature between sunshine and shade and air and water ... will be sufficient to produce results, which have hitherto been obtained only by a great expenditure of fuel'.

In 1823 the Commissioners of the Navy Board consulted Davy about the rapid decay of the copper sheathing of his Majesty's ships of war. A Committee of the Royal Society was appointed to consider the problem and Davy started to investigate it experimentally. He first showed that the corrosion was independent of small impurities in the copper and then, after examining the products of corrosion, he decided that it must depend on the oxygen dissolved in the sea water. Experiments having verified this conclusion, it occurred to him, in the light of his early researches, that he might prevent the oxidation of the copper by changing its electrical condition so as to make it slightly negative. It was not possible to do this in ships with a voltaic battery but it might be done by contact with zinc, tin, or iron. Laboratory experiments with zinc and iron in sea water gave perfect protection and large-scale trials gave similar results, so that the problem seemed to be solved. Davy then went on a voyage in the North Sea to measure the wastage of copper plates armed with zinc and iron protectors and some trials with ocean-going ships seemed to be successful. Unfortunately, it was then found that while the corrosion of the copper was prevented, the ship's bottom became so foul, from the adhesion of shells and weed, that her speed was greatly reduced. The Admiralty ordered the protectors to be removed just after Davy had read a paper to the Royal Society announcing the complete success of his plan.

The last paper Davy read to the Society was in 1826 on 'The relations of electrical and chemical changes', the subject of his first Bakerian Lecture twenty years earlier. It was mainly a restatement of the position and added little. In it he admitted the failure of his electrochemical protection of copper sheathing due to fouling, but suggested that the same principle, our modern cathodic protection, should be adopted to preserve the iron boilers of steam engines against corrosion. In one notable paragraph Davy foreshadows Faraday's classic determination of electrochemical equivalents: 'In the Bakerian Lecture of 1806, I proposed the electrical powers, or the forces required to disunite the elements of bodies, as a test or measure of the intensity of chemical union. By the use of the multiplier, it would now be easy to apply this test; and accurate researches on the connexion of what may be called the electro-dynamic relation of bodies to their combining masses or proportional numbers, will be the first step towards fixing chemistry on the permanent foundation of the mathematical sciences.'

I feel sure that Faraday, consciously or unconsciously, owed much to these flashes of Davy's genius. Faraday with his skill in measurement, his patience, and his unerring intuition gave precision and finality to Davy's tentative idea.

ILLNESS AND FINAL JOURNEYS

Davy's health was showing signs of failing before the great disappointment of the Admiralty experiment. The attacks that were made on him in the Press embittered him and the strain had serious results. He was just able to preside at the St Andrew's Day meeting in 1826, but shortly after he had a stroke and in January 1828 he went abroad with his brother first to Ravenna and then to his old haunts in Illyria, Upper Austria, and Bavaria, shooting and fishing whenever possible. In June he resigned the Presidency of the Royal Society and was succeeded by his early benefactor, Davis Gilbert. In October he returned to England.

Davy from his boyhood had had a passion for fishing. He was, as Maria Edgeworth said, 'a little mad about it'. Few men can have fished so many of the famous waters both in this country and on the continent and Davy took his fishing very seriously. He had made a scientific study of the structure and the habits of the Salmonidae and the best means of catching them. Lockhart in his life of Scott gives a description of Davy's costume at a sporting party at Abbotsford: 'The most picturesque figure was the illustrious inventor of the safety lamp. He had come for his favourite sport of angling . . . and his fisherman's costume—a brown hat with flexible brims, surrounded with line upon line, and innumerable fly-hooks; jackboots worthy of a Dutch smuggler, and a fustian surtout dabbled with blood of salmon—made a true contrast to the smart jackets, white cord breeches and polished jockey boots of the less distinguished cavaliers about him. Dr Wollaston was in black, and with his noble serene dignity of countenance might have passed for a sporting archbishop.'

The months after his return were spent mostly with his friend Mr Poole in Somersetshire and there Davy occupied himself with writing *Salmonia or Days of fly-fishing*, a book which occupies a place in angling classics between Izaak Walton's *Compleat angler* and Edward Grey's *Fly-fishing*. It is written in the conversational form and discursive style of the *Compleat angler*, for which Davy had a great admiration. Sir Walter Scott, also a fisherman, wrote a delightful appreciation of *Salmonia* in the *Quarterly Review*, comparing the sophisticated Davy with simple Izaak bottom fishing in the streams near London. The charm of Davy's book lies in his discussions, his theories, his wanderings in the byways of philosophy, of natural history, of religion—in fact his writing round fishing rather than of it. A description of the methods of carrying and preserving fish in Austria-Hungary leads him to enlarge on the virtues of the people and to condemn popular education and the King Press in England. He thinks wading as dangerous as hard drinking, he discourses on omens, he explains the colour of water, the virtues of crimping, the reasons of a belief in sea serpents and mermaids, and how Charles James Fox won a bet from one of the Royal Dukes counting the cats in St James Street. Davy was a very close and accurate observer of nature and many of the theories he evolved as the result of his observations have since been proved correct—that eels breed in the sea and not in rivers, and the migration of woodcocks. Writing *Salmonia* must have been a great solace to him.

When he was in London in January, 1828, he wrote to Knight: 'The garden of the Zoological Society is flourishing and there are a good many animals collected there.'

In March 1828 Davy set out on his final journey with young Dr Tobin seeking the *Consolations of a philsospher in travel*—the title of the book he wrote and left unfinished. Lady Davy remained in London entertaining her fashionable friends. They wandered for seven months through Styria and Illyria. 'I am contented and pleased with my little bantling,' Davy wrote from Gmunden when *Salmonia* reached him, 'more perhaps than I ought to be.' Its immediate success was a great pleasure to him. 'It has almost rekindled my love of praise,' he wrote from Laybach where he finished the corrections for the second edition in September. They reached Rome in November, where he wrote some chapters of his last book and continued some experiments on the torpedo, the electric eel. Lady Davy joined him at long last in April and they started home by easy stages reaching Geneva on 28 May 1829, where Davy died peacefully on the following morning and where he was buried by his own wish.

THE END

It was a sad ending to that brilliant life. Berzelius once wrote to Wöhler, when he had complained of his heavy task in translating the *Annual Reports* into German: 'The Herr Professor complains of so much writing. Yes, it's boring, but let's be clear that without it we could not do our best. If, for example, Davy had had to write in his youth as the Herr Professor has to now, I am convinced that he would have advanced chemistry by a whole century. But as it was, he left only brilliant fragments, just because he was never made to work diligently in all parts of chemistry as a whole.'

Berzelius was right. If Davy had had the disciplined training of Lavoisier, or if, instead of being apprenticed to Borlase, he had gone to Edinburgh in 1795 to have studied under Black, what might he not have accomplished? He had the imagination, he had the penetrating clearness of vision that is the mark of genius, he had the gift of experiment and the quickness of perception that always saw the next move, but he lacked training and he was often led astray by faulty measurements.

The early years in Bristol showed Davy's power when it was concentrated in continuous effort in one field. The Royal Institution with its distractions of agriculture and geology and the claims of society diffused his effort and led to short hours in the laboratory, often given to hasty and impulsive work. Davy's genius lay in chemistry and his incursions into other subjects were less fruitful. He was the brilliant pioneer of electro-chemistry, he exposed the myth of oxygen as the constituent of all acids, he foresaw the grouping of the elements in families, he discovered the principle of the safety lamp—on those his fame will rest. In wider fields his eloquence and enthusiasm did much to win public interest for science and he was quick to see how it could be put to useful purpose. He saved the Royal Institution at a critical moment of its early life and gave it the position it has held ever since. And in addition, he gave Faraday his chance, with all that was to mean to British science.

Fortune had smiled on Davy, perhaps too kindly in his younger years, and left him eager for praise, jealous of rivals, and anxious to shine in every field. Those were his failings, but with all his romantic genius made an enduring mark. We can leave him with the epitaph Berzelius wrote. When after Davy's death he had tied up, with sadness and regrets, the slender bundle of their broken correspondence, he wrote upon it, 'sitt tidehvarfs störste chemist'—the greatest chemist of his time.

5

THE PLACE OF
JÖNS JAKOB BERZELIUS
(1779–1848)
IN THE HISTORY OF CHEMISTRY†

WE ARE here today to do honour to a great son of Sweden, JÖNS JAKOB BERZELIUS, whose outstanding position in the first half of the nineteenth century made Stockholm a place of pilgrimage for young chemists from many countries. It is my privilege to pay a tribute to his memory by trying to describe for you the unique position he occupies in the history of chemistry. And what better setting could there be than this Conference Room of the Royal Swedish Academy of Science, surrounded by the portraits of Swedish men of science with Berzelius as the central figure. Facing me are the portraits of two of the three great scientists whose traditions Berzelius inherited—Linnaeus and Torbern Bergman—Scheele, alas is missing—and there too is the picture of that good friend Johan Gottlieb Gahn, from whom Berzelius learnt about the design of balances, the use of the blowpipe and much else. And adjoining us is the Berzelius Museum where his apparatus, his balances, and his preparations have been preserved with such loving care, and where one can almost sense the atmosphere in which he worked. I know of no equal memorial to the memory of another chemist.

I remember very clearly a day at school, more than fifty years ago, when I asked my chemistry master H. B. Baker, 'dry reaction' Baker, a question. I asked him 'Who was the greatest chemist in the nineteenth century?' Baker answered at once 'Berzelius', and I was frankly puzzled as I didn't associate him

† Lecture delivered at a meeting of the Swedish Royal Academy of Sciences, 22 September 1948.

Jöns Jakob Berzelius

like Lavoisier, Faraday, and Pasteur, with those dramatic episodes that catch the schoolboy's fancy. But it set me thinking and reading, and ever since Berzelius has been one of my good companions. And now thanks to the fine scholarship and devotion with which Dr Söderbaum has edited his letters and diaries and to his monumental biography we can know Berzelius much more intimately than was possible from his published works alone.

John Masefield, our British Poet Laureate in his poem on Biography, speaks of 'the bright moments of the sprinkled seeds'. Berzelius had those bright moments, and I will try to describe some of those episodes, when his mind responded quickly to the flash of inspiration, which had such a decisive influence on his life's work.

Berzelius was born in 1779, and during his boyhood Lavoisier was winning single-handed his victory over phlogiston by the use of the balance and by the principle of the balance sheet—Lavoisier wrote the first chemical equation. So that when Berzelius came to learn chemistry from Girtanner's *Elements of Antiphlogistic Chemistry*, that swift revolution had taken place that changed so profoundly not only chemistry but all the related sciences. And to Berzelius' logical brain the new system must have seemed so obvious that I sometimes wonder whether he quite appreciated how much Lavoisier had achieved in changing men's scientific outlook. But Berzelius had a sharp reminder that the advantages of the new system were not equally obvious to everyone when his first chemical paper was refused publication by the Swedish Royal Academy of Sciences in 1801 on the grounds that the Academy had not approved the new chemical nomenclature.

Berzelius was fortunate in his background. From his stepfather Ekmarck, from his boyhood friend Hagert and his schoolmaster Hornstedt he inherited the love of Linné for natural history and had his early training in systematics. At Uppsala, his University, there must have lingered the quantitative tradition of Torbern Bergman, that great analyst, and there too he must have learned first of Scheele for whom he had such unbounded admiration, Scheele, whose unerring eye and memory made him the greatest qualitative chemist of his generation. The strength of Berzelius lay in his almost unique combination of qualities. He was a great systematist, he saw chemistry as a whole. His approach to every problem, like Lavoisier's, was primarily quantitative. He was a fine craftsman and we owe to him much of our laboratory technique. But he had too the quick eye and the encyclopaedic memory that recognize immediately some new phenomenon, and he was the discoverer of three new elements and many compounds. In this he was a curious contrast to Lavoisier who lacked the qualitative sense and let the discovery of new compounds slip through his fingers. Fortunately for chemistry his brilliant contemporaries, Priestley and Cavendish, made the discoveries he needed, but Lavoisier alone saw their interpretation.

As a systematist Berzelius was the natural successor to Lavoisier and completed much that Lavoisier had begun. He too had his brilliant contemporaries, Davy, Dalton, Wollaston, Berthollet, Gay-Lussac, Thénard, and Dulong and many others who, following the new impetus that Lavoisier had given to chemistry, were making great contributions in their different fields. But it is mainly to Berzelius that we owe the integration of all that new knowledge into a coherent system.

Liebig in 1833 when telling Berzelius of his own work on various organic substances ended with the words, 'You see I am only hewing blocks of stone, and it must be left to you, our first architect, to erect the building and to design the plan.' Later Wöhler, at the beginning of the sad estrangement between them, reproached Liebig for his rudeness to Berzelius, 'who has done so much for the advancement of science and whose work has laid the foundations for its future development, with which the younger generation is now busy'.

Those words of these younger men describe so well the massive part that Berzelius played in the days when chemistry was becoming an independent science and was shaking off the trammels of medicine and pharmacy. Berzelius devoted his life to building the structure of chemistry. Like an architect he needed his foundations and elements of design, and when he had once decided on them he was reluctant to change lest the stability of his structure should be impaired. In 1814 in a letter to Thomas Thomson he says: 'It is clear that the theory of chemical proportions is intimately connected with the general theory of chemistry and that it forms the principal part of that

Jöns Jakob Berzelius

theory. But those in turn who are endeavouring to verify the theory, to analyse it or to extend it must keep their eyes fixed on the whole of chemistry. They must not adopt any theoretical view until they have traversed with it the whole terrain of chemistry, and before they have seen that it does not contradict any other part of the theory which there is good reason to accept as established.'

That I think explains Berzelius's outlook and makes it easier to understand the contrast between his obstinate conservatism at times and his genius for grasping new generalizations.

There is one other circumstance to be borne in mind. Berzelius had a hard upbringing and he knew little of the pleasures of childhood. From his earliest years he had to face the hard facts of life that left him little time for daydreams. And that I fancy left its mark on his outlook. His intellectual approach to any problem was factual. He was a stern critic of inaccuracy or wishful thinking. He spared no pains to gain the truth, and his standard of factual accuracy was one of his great contributions to chemistry.

During his medical studies at Uppsala Berzelius taught himself chemistry, and in the subjects he chose for his two university dissertations, a quantitative analysis of the mineral water of Medevi and the effect of the galvanic current on the human body, he gained his first experience of investigation in the two fields which were to dominate his life's work.

In 1802 Berzelius came to Stockholm, and during those early and somewhat precarious years here he was fortunate in the friendship and help of Hisinger and Gahn. In those formative years his researches covered a wide field: the classic investigation with Hisinger of the decomposition of salt solutions by the electric current, the use with Pontin of a mercury electrode with which he discovered ammonium amalgam and suggested to Davy the means of isolating calcium and barium, the discovery with Hisinger of cerium, the analyses of various minerals, and his first investigations of organic substances. Thus Berzelius was gaining that experience of a wide range of techniques that was a great source of his success as an investigator. But there was an amazing maturity in his early work. He always seemed to have his objective clearly in view, and it was the effectiveness of his attack and his economy of time and effort which made possible his vast output. You may remember Faraday's reply to a young man who asked him the secret of his success. Faraday answered, 'The secret is comprised in three words—Work, Finish, Publish', and there was the same incisiveness in Berzelius's method. His strength lay also in the thoroughness and speed with which he could master the literature of any subject. In his first book on galvanism, the prelude to his work on the effect of electricity on chemical compounds, he summarized in 1802 the advances in the knowledge of the electric current since Volta's and Galvani's discoveries. Dissatisfied with the teaching of physiological chemistry when he was a student he reviewed the whole subject in his lectures on animal chemistry published in 1806 and 1808. Many years later Wöhler was amazed at the amount of original work done by Berzelius himself which the lectures contained.

His output in those early years in Stockholm shows the grasp that Berzelius already had of problems in widely different fields but he says in his autobiography that his main interest would have been animal chemistry, but for the need of his students for a textbook of chemistry in their own language. With his usual thoroughness Berzelius began with a survey of the literature, and the decisive moment for his life's work came in 1807 when he was reading Richter's researches in stoichiometry. Richter, starting from the fact that solutions of neutral salts remain neutral after they have reacted, deduced that the amounts of different bases that neutralize the same quantities of different acids must be proportional to one another. He had shown also that the amount of oxygen is probably the same in the weights of different metallic oxides that neutralize the same quantity of acid. Berzelius saw at once the far-reaching consequences of Richter's ideas and the possibility of using them to analyse ammonium amalgam. He began by testing the accuracy of Richter's conclusions by the analytical results of Klaproth, Buchholz, and Rose. But as he says in a letter to Berthollet, 'After a year of almost fruitless work, I made up my mind at last to repeat all the analyses; I discovered basic defects and I arrived at a complete confirmation that I had at first rejected by obtaining always experimental results in agreement with my calculations!'

Berzelius's systematic mind saw the great opportunity, if

Jöns Jakob Berzelius

Richter was right, of calculating the composition of all salts and oxides from a limited number of analyses. He therefore set himself the task of improving the technique of quantitative analyses so as to reach the standard of accuracy required to make such calculations with certainty. It was a difficult task with the slender resources of his laboratory, and there was only one platinum vessel in Stockholm, too heavy to weigh on a chemical balance. But Berzelius had just the qualities that were needed—outstanding experimental skill and ingenuity, fine judgement in the choice of methods, great perseverance and industry, and a critical faculty that was hard to satisfy.

The work had not proceeded far when Berzelius read in *Nicholson's Journal* for November 1808 Wollaston's paper on Superacid and Subacid Salts. In this he learnt of Dalton's stroke of genius in giving the atomic theory quantitative significance by this Theory of Combination in Multiple Proportions. The critical sentence in Wollaston's paper was: 'All the facts that I had described are but particular instances of the more general observation of Mr Dalton, that in all cases the simple elements of bodies are disposed to unite atom to atom singly, or if either is in excess, it exceeds by a ratio to be expressed by some simple multiples of the number of its atom.'

Berzelius's quick mind must have seen in a flash the immense significance to chemistry of Dalton's generalization if it was true. Of this he soon found ample confirmation in the results he had already collected, and, as he said, 'the scope of my work, dealing originally with a very limited aspect of chemical proportions, was gradually enlarged until it finally embraced the whole field of chemical proportions, of which I had no true conception when I began my experiment'. The scope of Berzelius's work now widened into 'an attempt to determine the definite and simple proportions in which the constituent parts of inorganic substances are united with each other'. Berzelius devoted all his skill and industry to establishing the factual accuracy of Dalton's theory. He assumed nothing, and proved for instance that the proportions of lead and sulphur in lead sulphide and in lead sulphate are in fact the same. This he saw was fundamental to Dalton's conception.

Berzelius was eager to read Dalton's *New System*, but it was not until 1812 that he received a copy from Dalton himself. 'No

present', he wrote to Gaultier de Claubry, 'has ever given me so much pleasure as this did at first. But I will not conceal that I am surprised to see how much the author has disappointed my expectations. . . . In the purely chemical part his discrepancies from the truth rightly surprise me; one sees how he seeks to model nature to fit his hypothesis. His greatest fault is perhaps to prove nothing in a satisfactory manner; for this reason he cannot reveal to us the composition of fluoric and muriatic acids.' (Dalton had suggested that these were both compounds of hydrogen and oxygen.) . . . 'As for the experiments which are his own many seem to have been well done, but one cannot trust them, as one sees always that the author has a preconceived idea, and I know only too well from my own experience how often one is moved to determine in advance the result of an experiment by which one hopes to prove some theory!'

No sentences of Berzelius reveal more clearly his attitude of mind and his rigid factual approach. But in spite of his criticism Berzelius appreciated what chemistry owed to Dalton. In his 1810 paper he said: 'If Dalton's Hypothesis of Combination in Multiple Proportions is found to be true it will throw such new light on the theory of chemical affinity, that it will be the greatest advance towards the perfection of chemistry as a science, that has yet been made.' It was this conviction that lay behind those ten years of almost ceaseless work in the laboratory when Berzelius was laying the foundations of atomic chemistry.

And during those years Berzelius was constantly turning over in his mind the implications of the atomic theory and the ways in which its results could help building the structure of the science. First came the problem of affinity. Why did atoms combine, whence came the heat and light that so often accompanied their combination? Berzelius first proved that Lavoisier's theory that the heat of combustion came from the latent heat of the reactants was untenable. His mind then turned to his own experiments and those of Davy in which compounds had been decomposed into their constituents by means of electricity, and to the generation of light and heat in an electric discharge. He quickly reached a very simple conclusion: 'that in every chemical reaction, there is a neutralization of opposing electricities, and that this neutralization produces heat just as it is produced by discharges of the Leyden jar, of the electric pole or by thunder.' His instinct was right but he was left with the difficult problem of explaining the nature of the relationship between the electric charges and the atoms, and a century had to pass before the discoveries of the electron and the nucleus made this clear. Berzelius, however, was satisfied with the basic truth of the connection between affinity and charged atoms, and the electrochemical theory became one of his main articles of belief. What it meant to him can be seen from a sentence he wrote in the first flush of enthusiasm in 1814: 'Through the influence of electricity on the theory of chemistry, this last science has experienced a revolution, and received a greater and more important accession of influence than it did through the doctrines of either Stahl or Lavoisier.'

Alongside his electrochemical theory stood Berzelius's conception of dualism which he inherited from Lavoisier, but which received a new sanction from his views on the electrical nature of affinity. Oxygen for Berzelius as for Lavoisier was the central element. Salts were composed of basic and acidic oxides, and could be decomposed by the electric current into their electropositive and electronegative constituents. Similarly the basic oxides could be decomposed into their elements. Oxygen stood at the head of Berzelius's electrochemical series as the most electronegative of all elements, and therefore the only element whose electrical relations were invariable, as other elements could be electropositive to those higher in the list and electronegative to those below them.

But the importance of oxygen to Berzelius was not merely its electronegative character. His critical analysis of his quantitative results all centred on oxygen ratios in different oxides and in the basic and acidic constituents of salts. And it was on these oxygen ratios that he relied to determine the atomic composition of compounds. Berzelius had an instinctive belief in the simplicity of nature, and so in these early days he assumed that one element was always present as a single atom. From the oxygen ratios in the oxides of sulphur and iron he therefore assigned to them the formulae SO_2, SO_3 and FeO_2 and FeO_3. But having found that there are simple ratios also between the oxygen in the basic and acidic parts of salts, he used these ratios as well to check his results. For instance he found that the acidic oxide in neutral sulphates always contained three times as much oxygen as the basic oxide, thus confirming his formula SO_3. He found the same simple relationship of the oxygen ratio for double salts and for water of crystallization. And even when other methods were available, to the end of his life these oxygen ratios and to a lesser extent the corresponding sulphur ratios were his final criteria in fixing atomic composition.

Dualism was thus of immense practical significance to Berzelius, as it was the basis of his methods for determining atomic weights. Hence it is easy to see why he clung so obstinately to this point of view, and why he was so reluctant to admit the truth of Davy's view that chlorine is an element and not the oxide of an unknown radical.

Berzelius also made great use of Gay-Lussac's discovery that gaseous substances unite in simple proportions by volume. 'From what we know respecting definite proportions', he wrote in 1814, 'it follows that the combination in simple proportions by volume would hold with all bodies at temperatures at which they would assume the gaseous form. It is clear that what is in our theory called an *atom* is in the other a *volume*.' Berzelius used Gay-Lussac's law to assign formulae to the compounds of the permanent gases, such as H_2O to water, and in his first table of atomic weights he calls them 'specific weights'.

Following on his reform of Swedish pharmaceutical nomenclature he proposed a revised system of chemical nomenclature, using Latin names. As a sequel to this he introduced the chemical symbols we use today, in which an atom of an element is represented by the initial letter or letters of its Latin name. And thus we owe to Berzelius the chemist's shorthand.

His earlier quantitative work was all concerned with inorganic compounds, but with his interest in living things he was soon trying to ascertain whether organic compounds were subject to the same simple rules. It was not easy, and in January 1814 he wrote to Marcet saying that after working for three months without any sign of a result he nearly abandoned the attempt, but finally he succeeded in analysing fourteen organic acids by the combustion of their metallic salts. He found the compound atoms, as he called them, of organic substances much more complex than the inorganic but subject to the same quantitative laws.

By 1818 the whole field had been covered. Berzelius had

Jöns Jakob Berzelius

determined singlehanded the atomic weights of almost all the known elements. From these values and from his analytical results he assigned atomic formulae to some 2 000 compounds. It was an amazing achievement considering the limited resources of his laboratory and the state of chemical knowledge. There have been three classic investigators in this field, Berzelius, Jean Servais Stas, and Theodore Richards. Berzelius was the pioneer; greater refinements have come with modern technique, but the principles on which he worked still remain.

There were many factors that contributed to Berzelius's success: his critical study of the work of his prececessors; his encyclopaedic knowledge of the properties of chemical compounds and his experimental judgement in the selection of the most suitable reactions for his purpose; the skill with which he prepared his materials and the tests that he applied to assure himself of their purity. But above all his fine technique, the effectiveness and speed with which he worked, and the standard of accuracy which he set himself. Berzelius was a great craftsman, and he had the joy of the craftsman in fine tools. In 1816 he wrote to Hisinger: 'I've just got a delicious platinum evaporating basin from England, holding more than 300 cc. It's a jewel.'

In these ten years too Berzelius was laying the foundations of the modern technique of qualitative and quantitative analysis, which his students learnt from him and handed on. We owe to him so many of the simple laboratory devices and methods we still use today including rubber connections, the washbottle, the desiccator, greasing the edge of a beaker to avoid loss on transference, and lastly the perfection to which he brought the method of separating a precipitate by filtration, and his low ash filter papers which were the basis of a Swedish industry. His methods are described in the third volume of his *Textbook*, which has the same personal touch as Faraday's *Chemical Manipulation*.

The results of these ten years' work were summarized in his *Essay on the Theory of Chemical Proportions* which is an outstanding example of his wide grasp of the whole field of chemistry and of his clarity of exposition. It gives the evidence on which he based his atomic weight of each element and most of his values are very close to those we use today, except that, as he assumed a simple formula for metallic oxides, such as FeO_3 for ferric oxide, the atomic weights of most metals had twice their true value, and those of the alkali metals and silver were four times as large. The volume ends with a table giving the atomic formulae and the percentage composition of most of the known inorganic substances. It was the realization of the new outlook which Dalton had given to chemistry. Coming just at the time when the body of chemical knowledge was expanding so rapidly it was of inestimable value to chemists to have available the formulae of almost all the known inorganic substances with all the light they threw on the relation of one compound to another. Those tables are one of the milestones in the history of chemistry.

In the same year, 1819, came a dramatic discovery of great importance to Berzelius. Eilhard Mitscherlich, a young student, was then working in Berlin on the composition of the salts of phosphorus and arsenic. Berzelius had been surprised to find that the proportion of oxygen both in phosphoric and phosphorous acids and in the corresponding arsenic acids was 3 : 5, and had once suggested that the ratio might be the simpler one of 2 : 3. Mitscherlich had chosen for his thesis the task of ascertaining which of these two ratios was correct. In the course of his experiments he noticed that arsenates and phosphates of similar composition appeared to have the same crystalline form. Happening to meet Gustav Rose, a young student of mineralogy, at an evening party, Mitscherlich told him what he had discovered and they talked the whole evening about it, not even sitting down to supper with the rest. Next morning Rose borrowed the only reflecting goniometer in Berlin. They set it up together, following the instructions in Biot's *Physics*, and measured the angles of potassium dihydrogen phosphate and arsenate. They found them almost identical, as Mitscherlich had supposed, and so the Law of Isomorphism was discovered. It was diametrically opposed to Haüy's view which was generally accepted that each substance has a characteristic crystalline form. Mitscherlich and Rose joined forces, Rose helped in the measurement of the crystals and in return Mitscherlich taught him mineral analysis.

In the summer Rose was on a holiday in the Erzgebirge, and hearing that Berzelius was in Dresden on his return journey from Paris he went to see him and told him of Mitscherlich's discovery. By a coincidence Berzelius in a footnote in his 1819 *Essay* had speculated on the relations of crystalline form and composition and the conditions under which calcium, strontium and barium salts might have the same form, so the idea was not unfamiliar to him. He promised to see Mitscherlich and they met in Berlin shortly afterwards. Berzelius was immediately impressed by Mitscherlich's work and saw at once its immense significance in mineralogy and the simplicity that came from the idea of the isomorphous replacement of one element by another. Klaproth had died in 1817 and his chair was still vacant. Berzelius himself had refused it, and he now recommended to the Minister of Education that Mitscherlich should be the successor. The Minister made a counter suggestion that Mitscherlich should go to Stockholm to work in Berzelius's laboratory, and it was here that his classic paper was completed and the word isomorphism was coined. It was here too that Mitscherlich made his further discovery of dimorphism and in 1824 he returned to Berlin as Klaproth's successor.

Berzelius saw that Mitscherlich's discovery gave him a new means of comparing the atomic composition of isomorphous series of compounds. It was in fact the major factor in certain changes in his revised table of atomic weights that appeared in 1826. The isomorphism of the sulphates and chromates left no doubt in his mind that the formula of anhydrous chromic acid was CrO_3, corresponding to SO_3, and therefore chromic oxide must be Cr_2O_3. This led him to accept the more complex formulae such as Fe_2O_3 for ferric oxide and to halve the atomic weights of the metals, thus giving them their correct values except for six rare metals, the alkali metals and silver.

It is true that the discovery by Dulong and Petit in 1819 of the identity of atomic heats gave Berzelius another possible means of checking his atomic weights based on oxygen ratios. But there were exceptions and Berzelius never used Dulong and Petit's law with the same confidence as isomorphism.

For the permanent gases Berzelius had relied on Gay-Lussac's Law of Volumes and the determination of vapour densities by Dumas and by Mitscherlich opened up further possibilities of applying volume relationships. But the anomalies that occurred with sulphur, phosphorus, and mercury made Berzelius unwilling to accept this evidence. Progress in this field was dependent on the acceptance of the distinction made by Avogadro, by Ampère, and later by Gaudin between the atoms and the molecules of elements, and of the atomic complexity of their molecules. This Berzelius in common with other chemists would not then accept. It conflicted perhaps with the simple view he took of chemical composition, and he hardly ever used the word molecule, but spoke always of compound atoms.

Nature was in fact more complex in her atomic structure than Berzelius believed, and it was not until thirty years later that the full consequences of the logical application of these various physical methods of obtaining atomic and molecular weights were pointed out by Cannizzaro. As William Odling, who was in the heat of the battle of the fifty's, so often told me, chemists would not accept physical methods of molecular weight determination until their findings were confirmed by purely chemical evidence such as Williamson's work on etherification.

But in spite of his simple approach to the problem, Berzelius in 1835, by his sagacious use of alternative methods, assigned the correct relative magnitudes to the atomic weights of forty-two of the fifty-three known elements.

The completion of his ten years of concentrated work in the laboratory on atomic weights coincided very nearly with the election of Berzelius as Secretary of the Royal Swedish Academy of Science in 1818, and from that time he entered into a new sphere of activity.

One of his new duties, undertaken mainly on his own initiative, was his *Annual Report to the Academy on the Progress of the Physical Sciences*. With the rapidly growing output of research Berzelius saw the need for an annual review of scientific literature, and with his wide knowledge and experience none could have been better qualified to undertake it. It was a heavy task which Berzelius discharged with his usual thoroughness and clarity, severely critical of any work which did not come up to his standard of accuracy or objective approach.

The first slender volume appeared in 1822 with sections dealing with Physics and Inorganic Chemistry, Mineralogy, Animal and Vegetable Chemistry, and Geology. Year by year the *Reports* grew in size with greater emphasis on chemistry, and it is in them that one can follow best the train of Berzelius's thoughts in the second phase of his life.

In the first volume Berzelius wrote with enthusiasm about Mitscherlich's discovery of isomorphism which had been criticized severely by Haüy, the doyen of crystallography. Berzelius summed up: 'Haüy's contention is thus completely untenable. But one cannot expect that a scientist, towards the end of a life full of honour, should surrender, without resistance, without any attempt at defence, a position which he has regarded as the most important of his discoveries. That is perhaps more than one has a right to expect from any man.' I wonder whether Berzelius ever recalled that curious forecast of what his own position was to be in the bitter battles of the 1840s.

Once when the faithful Wöhler, who translated the *Reports* into German, had complained of his heavy task, Berzelius wrote to him: 'The Herr Professor complains of so much writing. Yes, it's boring, but let's be clear that without it we could not do our best. I curse the Annual Report when I begin it, but I praise it at the end when I see how much the work has added to my own stock of knowledge.'

Berzelius was right in the importance he attached to the *Reports*, which were invaluable to chemists at a time when knowledge was increasing so rapidly. He had always insisted on the need for a synoptic view of chemistry, and the *Reports* gave him the opportunity of surveying progress along the whole line of attack, and making from time to time those generalizations which have taken a permanent place in our chemical philosophy.

First came isomerism. By 1832 a number of instances were known of pairs of substances with the same composition but different properties—fulminic and cyanic acids, the two phosphoric acids, the oxides of tin, and racemic and tartaric acids, the last two observed by Berzelius himself. 'Since it is important', he wrote in the *Annual Report*, 'that accepted ideas should have definite and logical names, I propose to call substances with the same composition and different properties isomeric.'

In the following year he added the class of polymers to include substances of the same percentage composition but containing different numbers of atoms like ethylene and Faraday's butylene, as we call them now.

In 1835 he reviewed a number of reactions which take place in presence of some substance which remains unaffected, having apparently taken no part in the change: the transformation of starch to gum and then to grape sugar in presence of dilute acids, the decomposition of hydrogen peroxide in presence of acids or platinum black, the oxidation of alcohol in air in presence of platinum black, the etherification of alcohol in presence of sulphuric acid, and the effect of diastase on starch. Berzelius saw in these reactions some common cause, the nature of which was unknown, and he gave it the name catalysis. 'The catalytic power of substances', he wrote, 'seems to depend on their ability to awaken the dormant affinities of other substances by their mere presence and not by their own affinity.' A sentence in a letter to Liebig in 1835 reveals Berzelius's deep insight into living processes. 'I believe it is extremely probable', he says, 'that this catalytic force plays a major part in living nature, and that the many chemical changes in the secretory organs of animals and plants are caused entirely by substances on the inner coating of those organs which bring about the changes in the liquids passing through them.'

Berzelius all through his life kept a lively interest in physiological chemistry, and at this period was in constant correspondence with Mulder, a Dutch chemist working in this field. In 1838 Berzelius wrote to him, 'The name protein I suggest to you for the organic oxide of fibrine and albumen is derived from πρωτειος, because it appears to be the prime or principal substance in animal nutrition which plants produce for herbivora, from whom it passes finally to the carnivora':†

And so the words isomeric, polymeric, catalysis, and protein, which are so often on our lips, we owe to Berzelius, to the watchful eye he was keeping on the whole field of chemistry, to his genius for realizing when a generalization was justified, and for coining words, almost the inevitable words, which we use today to express almost the identical ideas that were in his mind.‡

Alongside his literary work on the *Annual Reports* went the completion of his great *Textbook*, the source book of all its successors. Volume by volume it appeared, covering the whole field, enriched often by his own experiments to verify some doubtful statement or to fill some gap in knowledge. For in

† See note on p. 152.   ‡ Ibid.

Jöns Jakob Berzelius

spite of all his other labours Berzelius was constantly in the laboratory and the results were shown in a series of papers covering the most diverse fields—the redetermination of atomic weights, the analysis of minerals and meteorites, comprehensive studies of the ferro-cyanides, the sulpho- and seleno-cyanides, and the fluorides, the chemistry of the platinum metals, of vanadium, molybdenum, and uranium, the discovery of thorium, improved methods of quantitative analysis, and the investigation of many naturally occurring organic substances including chlorophyll.

There we see Berzelius in Liebig's words as the architect. The foundations had been laid in his assignment of atomic formulae to almost every compound and now he was completing the structure in a self-effacing way—not 'chasing discoveries', but systematically filling in the gaps in knowledge wherever he found them.

And in those years successive generations of young chemists came to Stockholm to work in his laboratory to learn the secrets of technique which made his work so effective. Long before Liebig at Giessen, Berzelius had a school of chemists, the first of its kind.

And so for many years Berzelius was the accepted arbiter until in the late thirties clouds began to gather on the horizon and his authority was challenged. It was the prelude to the stormy battles of the forties when controversy in chemistry reached its most bitter phase and men suffered for their scientific beliefs.

The challenge to Berzelius came from his attitude to the newly discovered phenomenon of the substitution of hydrogen by chlorine in organic compounds, and from the growing opposition to his dualistic view of chemical composition.

The discovery by Dumas and by Laurent of organic compounds in which chlorine had replaced hydrogen, such as chloracetic acid and the chlornaphthalenes, without changing the general characteristics of the parent substance led to the view that an atom of chlorine could take the place of an atom of hydrogen. To Berzelius with his theory of affinity based on the electrical polarity of atoms it was unthinkable that an electro-negative atom of chlorine could take the place of an electro-positive atom of hydrogen. This made it necessary for him to assign a different formula to the chlorinated derivatives for which there was no experimental evidence, as for instance the chlornaphthalenes which Laurent had shown were generally isomorphous with naphthalene itself. Finally in 1840 Berzelius compromised by adopting Gerhardt's conception of bodies like nitrobenzene being conjugate compounds produced by the union of two residues with the elimination of water, the organic radical being called by Gerhardt a 'copula'. Berzelius accepted the substitution of hydrogen by chlorine in the copula and conjugate compounds and copulae became the order of the day. It was in fact his surrender of a position which new facts had made untenable and it had an interesting sequel. The young Hermann Kolbe, working in Wöhler's laboratory, published a paper on the chlorinated methyl sulphonic acids which he formulated as conjugate compounds with the radical methyl acting as a copula. Berzelius was delighted at the discovery of a series of analogues to the chloracetic acids. He devoted six pages of the *Annual Report* to Kolbe's paper and won his heart by writing him an encouraging letter which Kolbe kept as a talisman all his life. That success coloured the whole of Kolbe's work, and in his classic paper in 1860 on 'The Natural Relations between Organic and Inorganic Compounds' in which he predicted the existence of secondary and tertiary alcohols he was following Berzelius in the idea that the constitution of organic compounds must be derived from their inorganic analogues.

However, on dualism Berzelius could not compromise. That was the basis on which he had constructed his system of atomic weights and atomic formulae, so that any compromise must have seemed to him to undermine the whole structure which he had built up over so many years. But the work of Graham and of Liebig on polybasic acids had turned the thoughts of chemists to a unitary conception of chemical molecules. Dualistic formulae like that of Berzelius for acetic acid, $C_4H_6O_3.H_2O$, were no longer acceptable to the younger generation, and even his atomic weights were replaced by equivalents.

It was inevitable that the old landmarks should be swept away, and that with the loss of confidence in the old order there should come a return partly to empiricism and partly to the individual theories which were the basis of each man's contribu-

Jöns Jakob Berzelius

tion. It was no wonder that by 1859 a whole page in Kekulé's textbook was filled by the different formulae proposed for such a simple substance as acetic acid. It must have been a frustrating experience for Berzelius, who had striven all his life for order, but he never relaxed his efforts to contribute all he could to his beloved science.

He continued to write single-handed the whole of the *Annual Report on Chemistry and Geology*, Baron Wrede having taken over the physics section in 1839. At the same time Berzelius was re-editing the volumes of his *Textbook* with his usual care and thoroughness. His correspondence with the faithful Wöhler shows the keenness with which he was following every new development, and the way in which he re-examined any doubtful point himself in his laboratory.

In July 1847 he had a bad attack of gout and complained to Wöhler of his weakness. Wöhler replied that he was not altogether surprised as he feared that Berzelius's health might suffer from the immense amount of writing he had been doing. 'The Fourth and Fifth Volumes (of the *Textbook*) were a herculean task and done so quickly . . . and the *Annual Report* as well . . . It would be a calamity if you gave up the *Annual Report* as it is the governor, the guide, that keeps a crazy train on the track.' But Berzelius could write no more and in his last dictated letters to Wöhler his two anxieties were the continuation of the *Annual Report* and the completion of his *Textbook*, 'my legacy to the future . . . I commend it to you like a father's only child'. He died in August 1848.

Science never had a more devoted, a more loyal son than Berzelius. He gave his whole life to her service with great tenacity of purpose and effectiveness in execution. Chemistry was indeed fortunate that in the years when it was becoming an independent science and the stream of knowledge was growing so rapidly it had the encyclopaedic mind, the judgement, the craftsmanship, and the watchful eye of Berzelius to guide its career. It was that rare combination of qualities that made possible his massive contribution which is unique in the history of chemistry. His systematic mind saw the need for a structure in which chemistry could grow with the precision and the articulation of a living organism. The basic principle of his design was atomic composition. At a time when there were many differences of opinion and some scepticism, Berzelius made Dalton's theory effective by his pioneer work in determining atomic weights and atomic formulae. A more subtle and a less executive mind might well have been baffled by the complexities which took decades to unravel. Berzelius laid the foundations of a structure which was ultimately modified through the interpretation of new knowledge. The fixity of purpose and the simplicity of mind, which were his strength, made him reluctant to admit changes in his basic principles. And it is that reluctance which is sometimes remembered, when the immense debt we owe to him is forgotten. So much of his great legacy seems to us today so obvious, as Lavoisier's must have seemed to him. The accurate fixing of atomic weights, atomic formulae, a comprehensive system and the ideas lying behind the words he coined—isomeric, polymeric, catalysis, and protein—all those we owe to Berzelius. And behind his great accomplishment lay that mastery of technique that has become our heritage. Berzelius is one of the immortals in the history of chemistry.

Note (26.6.1949). Mulder is generally credited with having coined the word protein which first occurs in print in his paper 'Sur la composition de quelques substances animales' in the monthly issue of the *Bulletin des Sciences physiques et naturelles Néerlande* dated 30 July 1838 (vol. I, p. 111) in the following sentence: 'La matière organique, étant un principe général de toutes les parties constituantes du corps animal, et se trouvant comme nous verrons tantôt, dans le règne végétal, pourrait se nommer Protéine de πρωτειος primarius.' Berzelius's letter quoted on p. 148 was dated 10 July 1838, the postmark shows that it reached Greifswald on 14 July, and although the postmark at Rotterdam is illegible, the evidence of the transit time of another letter makes it probable that it reached Mulder by 20 July. Mulder's paper opens with a generous acknowledgement of his debt to Berzelius, and there seems to me little doubt that he adopted Berzelius's suggestions without making any specific acknowledgement to him.

6

MICHAEL FARADAY
(1791–1867)
AS A PHYSICAL CHEMIST†

IN speaking of Faraday in the Royal Institution my thoughts inevitably go back to the young bookbinder's apprentice sitting in the gallery of the Lecture Theatre over the clock taking those careful accurate notes of the last four lectures which Davy delivered here. Having read them, how Davy must have welcomed the opportunity, two months later, after Mr Payne, the laboratory assistant, had assaulted Mr Newman, the instrument maker, of recommending Faraday for the vacant post.

In his letters to Abbott at the age of nineteen we already get the first glimpse of Faraday's amazing natural gifts and his love of science when with only the rudiments of education he is describing his experiments in electricity and arguing convincingly about the true nature of chlorine. For twelve years Faraday helped Davy in his researches and he was always grateful for that experience. It was a happy chance that brought together two men with such genius for experiment.

At the start Davy is working on nitrogen chloride and fluoric acid and in October 1813 Faraday went with him on his first long foreign tour. In Paris he is helping in the work on

† Address at the Jubilee Meeting of the Faraday Society in 1953 with some extracts from the Presidential Address to Section B of the British Association in 1931, 'Michael Faraday and the Theory of Electrolytic Conduction'. The title of the lecture to the Faraday Society was chosen deliberately to show that Faraday was much more than an electro-chemist and that he was one of the pioneers of physical chemistry as shown by his researches on the liquefaction of gases, the viscosity of gases, musical flames, vapour pressures, the freezing point of water, steel alloys, his discovery of magneto-optical rotation and the range of his experimental techniques in *Chemical Manipulation*.

iodine, in Florence with the combustion of the diamond, and in Rome with the discovery of iodine pentoxide. Back at the Royal Institution, in October 1815, he is sharing in that brilliant fortnight's work that resulted in the safety lamp, and making a fair copy of the paper for the Royal Society. Then Davy starts him on his first independent investigation and there soon followed a steady flow of papers characterized by their accuracy and the careful avoidance of any theoretical commitments. They range over a wide field without any consistent theme, recording the results of Faraday's investigations of any new substance or phenomenon he encountered. 'On the Escape of Gases through Capillary Tubes (1817); 'On the Solution of Silver in Ammonia' (1818); 'On two new Compounds of Chlorine and Hydrogen' (1820), in which he isolates hexachlorethane and tetrachlorethylene; 'On the Condensation of Several Gases into Liquids' (1823); 'On new Compounds of Carbon and Hydrogen' (1825), recording the discovery of benzene and butylene; 'On the Mutual Action of Sulphuric Acid and Naphthalene' (1826), in which he separates the barium salts of α- and β-naphthalene sulphonic acids.

Faraday's outlook in this early work is shown so well by the final sentence of his paper on naphthalene sulphonic acid. He suggests for this the name 'sulpho-naphthalic acid, which sufficiently indicates its source and nature without the inconvenience of involving theoretical views'.

These investigations show Faraday's capacity as a practical chemist, the neatness and simplicity of his experimental methods, the quickness and accuracy of his observation, and the completeness with which he treated a subject. It was Faraday's good fortune that so important a substance as benzene was in the gas oil given to him by Gordon to investigate, but the remarkable part of the work is his separation of benzene by fractional distillation and crystallization, and the accuracy both of his analysis made with the simplest means and of his vapour density determinations made by exploding a known volume with oxygen and measuring the contraction and the volume of carbon dioxide formed.

In those years Faraday was gaining that first-hand acquaintance with the properties of many substances, which was to be invaluable to him in choosing the right materials for his experiments. He was also accumulating an unrivalled knowledge of chemical technique and gaining the confidence to which was due the boldness and directness of his experiments in the years to come.

His own patient detailed study of laboratory operations gave his book *Chemical Manipulation*, published in 1827, its unique character, with its descriptions of every kind of laboratory operation and device. The art of experimenting must almost necessarily be traditional. Faraday, however, wanted to make chemical manipulation less of an alchemical secret, 'taught only in the very depths of the laboratory to a highly privileged few'. The book is one of the most personal documents in scientific literature as each page is a record of his own experimental methods, showing how every detail of each operation had been thought out by him and reduced to its simplest and most effective form.

There is an intensively personal quality about Faraday's work, as it was all done with his own hands, and even if he used the results of others he repeated their experiments. 'I was never able to make a fact my own without seeing it. . . . If Grove, or Wheatstone, or Gassiot told me a new fact and wanted my opinion . . . I could never say anything until I had seen the fact. For the same reason, I never could work, as some professors do most extensively, by students or pupils. All the work had to be my own.' Faraday worked alone to the end of his life with no helper except the trusty Sergeant Anderson, who for almost forty years was his laboratory assistant. 'He and I are companions, in years, in work and in the Royal Institution.' Anderson deserves a place in our chemical hagiology beside Berzelius's faithful cook, Anna, whose conservatism as regards the nature of chlorine was even greater than her master's.

Throughout his life, Faraday had an intense interest in the applications of science to everyday problems, often making them the subjects of his Friday evening discourses at the Royal Institution. We are apt to forget that he was a skilful analyst and that for several years he made a considerable income as a consulting chemist by what he called his 'professional business', until in 1831 he deliberately gave up this work lest it should interfere with his researches.

His papers with Stodart, a maker of surgical instruments, on steel alloys are typical of his practical outlook, as the work is directed solely to the discovery of steels resistant to corrosion with superior cutting powers. Faraday would sometimes give his friends a razor blade made with his alloy steel. The investigation ceased with Stodart's death in 1822.

Government departments were constantly seeking his help and advice, and for thirty years he was Scientific Adviser to Trinity House, where he gave his time unsparingly to such problems as lighting and ventilation, and even to the examination of water supplies and of samples of oils and paints. Quite late in life he had not lost his cunning as an analyst. In 1845 he was reporting to Trinity House on the adulteration of white lead, and in 1852 he made an analysis for the Board of Ordnance in two days of the contents of a French shell fired at Salee.

In 1825 Faraday's help had been enlisted in the research sponsored by the Royal Society and the Board of Longitude to improve the quality of British optical glass: furnaces to his design were built at the Royal Institution and he interested himself in all the practical details of glass making. The only positive result of the work was the heavy lead borosilicate glass with which twenty years later he discovered the magnetic rotation of polarized light. It was Faraday's skill in applying scientific methods to practical use that made his services so much in demand by government departments throughout his life.

In his description of the requisites of a laboratory, Faraday wrote, 'A blank writing-paper book should be upon the table, with pen and ink, to enter immediately the notes of experiments. A chair may be admitted, and one will be found quite sufficient for all necessary purposes, for a laboratory is no place for persons who are not engaged in the operations going on there. . . . The practice of delaying to note until the end of a train of experiments or to the conclusion of a day, is a bad one, as it then becomes difficult accurately to remember the succession of events. There is a probability also that some important point which may suggest itself during the writing, cannot then be ascertained by reference to experiment, because of its occurrence to the mind at too late a period.'

Faraday's own notebooks are much more than a record of his experiments, as he jotted down in them in numbered paragraphs, which ran to 16 041, the ideas which flashed on him as he was working in the laboratory and his plans for new experiments, so that we can follow his progress from day to day and watch the interplay of ideas and experiments, and the swiftness and certainty with which he reached a decision. When a young man asked him the secret of his success as an investigator, Faraday answered, 'The secret is comprised in three words—Work, Finish, Publish'. What he meant by this we can see by following in his notebooks the course of his researches in electrochemistry from 1831 to 1834.

It is easy to see why Farady had to work alone with nobody to distract him. In the period of his great achievments, his experiments were rarely continuous, the intervals between them suggesting the subconscious working of his mind. He waited until the impulse came and his 'prescient wisdom' had planned the experiment and foreseen the result. As we read the pages of the notebooks, discovery seems to follow discovery almost inevitably. Faraday always had a preconceived idea behind his experiments, and never were advances made with such economy of effort. Each new position was reached by a series of attacks delivered with amazing speed when everything was ripe for them. The eager intensity with which Faraday worked in the laboratory impressed all those who watched him: 'His motions were wonderfully rapid; and if he had to cross the laboratory for anything, he did not walk at an ordinary step, he ran for it, and when he wanted anything he spoke quickly....' 'The rare ingenuity of his mind was ably seconded by his manipulative skill, while the quickness of his perceptions was equalled by the calm rapidity of his movements.'

Faraday followed Davy and Wollaston in their doubts about the atomic theory and consequently he had no coherent theory of the relation of different chemical substances to one another. He made no attempt to grapple with the problem of their constitution which was just beginning to excite the curiosity of chemists. The truth is that Faraday in spite of his many contributions to chemistry was by nature a physicist. During those years from 1816 to 1830 when his papers were entirely practical, eschewing theoretical inferences, his inner mind was brooding over his early conviction of the essential unity of nature, intent on finding the underlying relation of light, electricity, and magnetism. But thanks to his early training with

Davy, his wide acquaintance with the properties of many substances, and his unrivalled knowledge of chemical technique, there was always a chemist's background to his experiments which gave them wide scope. It was his dual outlook that made him a key figure in the early history of physical chemistry.

The year 1831 was the turning-point in Faraday's career. There is no greater contrast in scientific literature than his earlier chemical and physical papers characterized by their essentially practical outlook and accomplishment, and the brilliant flights of imagination that inspired his 'Experimental Researches in Electricity'.

I am convinced that it was the success of an experiment which Faraday had previously tried again and again, without success, that gave the new impulse to his work and gave him confidence in the promptings of his imagination.

Electricity was one of Faraday's earliest scientific interests. Long before he went to Davy he was experimenting with home-made batteries. In his first lecture to the City Philosophical Society in 1816 we get a glimpse of his intuitive belief in the essential unity of the forces of nature. 'That the attraction of aggregation and chemical affinity is actually the same as the attraction of gravitation and electrical attraction I will not positively affirm, but I believe they are.' In 1821, Faraday repeated the experiments of Oersted, Arago, and Ampère on electromagnetism and discovered the rotation of a wire carrying a current if free to move round a magnetic pole. Magnetism had been produced from electricity, and Faraday was convinced of the possibility of obtaining electricity from magnetism. In 1824 and again in 1825 and 1828 he was experimenting with a magnet in a wire helix connected with a galvanometer, without result. Either the galvanometer was too insensitive or he failed to detect the momentary deflection when the magnet was introduced. On 29 August 1831, the induced current was detected—Faraday's dream had come true—and ten days of decisive experiment ended in his paper on 'The Induction of Electric Currents' which was to shape the future of electrical science and electrical industry.

It is significant of Faraday's train of thought as a pioneer of physical chemistry that the first experiment he made on 29 August 1831, after discovering the induced current, was to attach platinum wires to the ends of the coil and see if he could detect any decomposition of a drop of copper sulphate solution. The test was not delicate enough, and he did not succeed in detecting the chemical power of magneto-electricity until several years later. However, he made a number of experiments on chemical means of detecting the current from a voltaic cell, and on 11 June 1832, he found in bibulous paper moistened with potassium iodide and starch the most sensitive detector.

His discovery of electromagnetic induction raised afresh in Faraday's mind the old and still disputed problem of the identity of electricities from different sources, and chemical action was one of the tests he applied to its solution. Having shown that common (frictional), voltaic, and magneto-electricity all produce similar physiological, magnetic, chemical, and thermal effects, Faraday sought to establish their identity by quantitative experiments. Among the discoveries that Cavendish had made almost fifty years previously, was that the behaviour of an electric charge depends on two factors, the degree of electrification (or potential as we should call it) and the size of the charge. His paper on the torpedo, published in 1776, contains an account of experiments with Leyden jars which demonstrate very clearly this distinction. Faraday had read Cavendish's paper and realized that this theory explained the apparent differences in behaviour of frictional and voltaic electricity. He says: 'The beautiful explication of these variations afforded by Cavendish's theory of quantity and intensity requires no support at present, as it is not supposed to be doubted.' The theory was, however, by no means generally accepted in 1832, but Faraday's recognition of its truth was the key to the success of his researches on electrochemistry.

He started out to prove that the effect on a galvanometer of a discharge of frictional electricity was dependent on its quantity and not on its intensity, which he showed on 14 September 1832, by charging eight Leyden jars in a battery of fifteen by means of thirty turns of a frictional machine, and seeing that when discharged through the galvanometer† they caused the

---

† Fortunately the galvanometer he used had ballistic properties and so measured the amount of electricity discharged through it. Faraday records its half period of swing and was careful to reduce the total time of discharge to half this value.

same throw of the needle as the whole fifteen jars charged by means of thirty turns, although an electrometer indicated that the intensity of the charge was roughly one-half in the second case. 'Hence,' said Faraday, 'it would seem that if the same absolute quantity of electricity passed through the instrument whatever may be its intensity, the deflecting force is the same.'

He then made a small 'standard elementary battery' consisting of platinum and zinc wires $\frac{5}{8}$ inch in diameter, $\frac{5}{16}$ inch apart, which he immersed to a depth of $\frac{1}{8}$ inch in dilute sulphuric acid (one drop of acid in 4 oz water) and he found that by immersing the wires for eight beats of his watch (3·2 s) when they were joined to a galvanometer, the momentary deflection of the needle (its half period of swing was 6·8 s) was the same as that caused by thirty turns of the machine. Hence, said Faraday, the amount of electricity produced by the cell in this time was the same as that produced by thirty turns of the machine. He next compared the amounts of chemical change by allowing the electricity to pass in both cases through filter paper moistened with potassium iodide on a platinum spatula with a platinum wire $\frac{1}{12}$ inch in diameter as the positive pole. A brown circle of iodine was found at the point of contact and its tint depended on the number of turns of the machine. Faraday showed by varying the number of turns of the machine that it required approximately thirty turns to produce an iodine spot of the same tint as that given by immersing his standard battery for eight beats of his watch. 'Hence it would appear that both in magnetic deflection and in chemical effect the current of the standard voltaic battery for 8 beats of the watch was equal to the electricity of 30 turns of the machine, and that therefore common and voltaic electricity are alike in all respects.' These experiments on 14 and 15 September 1832, were the first attempt to connect the quantity of electricity passing in an electrical circuit with the amount of chemical decomposition.

Faraday was now convinced, it is true on rather slender evidence, that the amount of electrochemical decomposition is a measure of the quantity of electricity, and his paper on 'The Identities of Electricities' contains the first statement of his First Law of Electrolysis: 'It also follows that for this case of electro-chemical decomposition and it is probable for all cases, that the chemical power, like the magnetic force, is in direct proportion to the absolute quantity of electricity which passes.'

In trying to find the most delicate test for the passage of electricity, Faraday made many new experiments on the chemical action produced by a current. In one of them he placed one end of a long piece of litmus paper moistened with sodium sulphate in contact with an electrical machine, whilst the other end was held opposite to the discharging points. On turning the machine Faraday saw that decomposition took place, the paper becoming red 'where the positive electricity entered from the air'. This proved to him that the decomposition was not dependent on the presence of metallic poles in the solution, and on 6 September he wrote in his notebook:

Hence it would seem that it is not a mere repulsion of the alkali and attraction of the acid by the positive pole, etc., but that as the current of electricity passes whether by metallic poles or not the elementary particles arrange themselves and that the alkali goes as far as it can with the current in one direction and the acid in the other. The metallic poles used appear to be mere terminations of the decomposable substance.

The effects of decomposition would seem rather to depend upon a relief of the chemical affinity in one direction and an exaltation of it on the other rather than to direct attraction and repulsion of the poles.

Faraday's view of the nature of electrolysis had been foreshadowed by Davy in his Bakerian Lecture of 1806 and I feel certain that in his electrochemical researches, consciously or unconsciously, Faraday owed much to Davy's flashes of inspiration. In his papers scattered over the years 1801 to 1826 Davy had touched on many aspects of the researches that were to occupy Faraday during the next eighteen months. And Faraday's skill in measurement, his patience, and his unerring intuition were to give precision and finality to Davy's tentative ideas.

The three main problems he now attacked were:

1. The mechanism of conduction in solution and the dependence of the passage of the current on chemical decomposition.

2. The amount of chemical action accompanying the passage of the current.
3. The source and the intensity of the current produced by voltaic cells.

On 24 December 1832, Faraday wrote in his notebook: 'Can an electric current, voltaic or not, decompose a solid body, ice etc., etc.? If it can does it give structure at the time? If it cannot what would fused gum, lac, wax, etc.?' A cold spell at the end of January enabled him to put this to the test, and he found that while ice would not conduct a voltaic current, conduction occurred immediately the ice melted. 'If ice will not conduct is it because it cannot decompose?'

It was characteristic of Faraday's thoroughness that he went on to examine the conductivity in the fused state of a number of substances which are solid at ordinary temperatures and to study the products formed during electrolysis. It was a new field for him and he showed his usual experimental skill in devising simple methods for working at high temperatures including even the use of the oxyhydrogen blowpipe. During February and April he examined over 130 substances and found that while a number resembled water in being insulators in the solid state and becoming good conductors if fused, when they were decomposed by the current, certain of them, such as boric acid, did not conduct when fused. He thus arrived at no general conclusion, but the experience he gained in working with fused salts was to prove invaluable later in the year in his work on electrochemical equivalents. The experiments were finished on 22 April, and on 24 April they were communicated to the Royal Society with the title 'On a New Law of Electric Conduction'.

Faraday then turned his attention to the mechanism of conduction in a solution. On 2 May he passes a strong current through a saturated solution of sodium sulphate and examines it with polarized light both across and along the direction of the current to see if he can detect signs of arrangement of the molecules, but without result. On 20 May he determines the transfer of sulphuric acid during electrolysis by measuring the changes in concentration in two vessels connected by moist asbestos, and on 27 May he shows that the transfer of sulphuric acid differs from that of sodium sulphate of equivalent concentration, 'very evident therefore that the transfer is dependent on the mutual action of the particles'. He summed up his views in a paper to the Royal Society on 18 June, the main conclusion being

'that electro-chemical decomposition does not depend on the simultaneous action of two metallic poles', and the effects of it are due to a modification, by the electric current, of the chemical affinity of the particles through or by which that current is passing, giving them the power of acting more forcibly in one direction than in another, and consequently making them travel by a series of successive decompositions and recompositions in opposite directions, and finally causing their expulsion or exclusion at the boundaries of the body under decomposition.

In May 1833 Faraday's thoughts were returning to the question of the amount of chemical action that takes place during the passage of a current. On 16 May no experiments were recorded in the notebook, but among the ideas he jotted down was:

Is the law this [above a certain intensity, i.e. the one required for decomposition to take place at all], that whatever the size of plates, or number intervening, or constant section of decomposing matter, or variable section, or variable strength, or number of series in the battery: that . . . equal currents of electricity measured by the galvanometer evolve equal volumes of gas or effect equal chemical action in a constant medium?

A week later he writes down his plans for testing the law. 'By putting cups and expts. in succession and sending the same electrical current through both or all am sure that each is submitted to an equal force. Can try well this way whether the same quantity of different intensity does the same chemical work using same dilute sulphuric acid but *different-sized poles*, and collecting gas, and that will tell—some poles mere wires, others large plates.' Three months elapsed before he actually carried out the experiment. On 27 August he wrote: 'Pursue the investigation, whether the same quantity of electricity *always* produces an equivalent of chemical decomposition. . . .' On 31 August he found, as he expected, that the same amount of current liberated the same volume of gas irrespective of the concentration of the acid, the size of the electrodes, or the intensity of the current. He obtained the same results with solutions of various salts, and his comment was: 'Strange that with such different substances the same quantity of water should be decomposed by the same current.' These experiments were continued in September, and Faraday was constantly puzzling over the effect of various substances in increasing the conducting power of water. On 17 September he showed that cells containing muriatic and sulphuric acids had given the same volume of hydrogen when connected in series, and he was now busy constructing a simple apparatus to measure the quantity of electricity by means of the volume of gas produced by it. 'The instrument offers the only actual measurer of voltaic electricity which we at present possess . . . I have therefore named it a volta-electrometer.' (The name was contracted to voltameter five years later.) Today, following Faraday, we define our practical unit of current by its electrolytic action, and we use his name to denote the fundamental unit of electrochemistry.

On 19 September, among his observations, he notes 'Will not white-hot diamond conduct? If so may perhaps crystallize carbon at white heat by power of the voltaic battery.'

He had been worried by the contraction, on standing, of the mixture of oxygen and hydrogen obtained in the electrolysis of sulphuric acid. He traced this to the catalytic activity of the platinum electrode, and showed that the positive and not the negative was effective. This observation led him to spend some weeks investigating the conditions under which platinum and other metals would assist the combination of various gases, when he discovered the retarding effects caused by small quantities of gases such as olefiant gas, carbonic oxide, and sulphuretted hydrogen. The results were communicated to the Royal Society on 30 November.

Faraday then returned to the investigation of the amount of chemical action produced by the current, and as he recognized that in the electrolysis of aqueous solutions it was doubtful whether the elements liberated at the poles were to be regarded as primary or secondary products, he extended the inquiry to include fused substances, which would be free from this ambiguity.

Hitherto Faraday had only compared the quantities of the

same substances, such as hydrogen, liberated by the same current in a series of cells, but he now began to consider the relative quantities of different elements that would be liberated by the same current. On 23 September he wrote: 'Think it will be very important to have a new relation of bodies, under the term *electro-chemical equivalents*, tabulated. Very important as to decomposing powers of the pile, as to the true expression of equivalent numbers, and as to nature of chemical affinity and its relation to electrical states and powers.'

In the last paper Davy read to the Royal Society in 1826 there is a notable passage in which he foreshadowed Faraday's classic investigation. 'In the Bakerian Lectures of 1806 I proposed the electrical powers, or the forces required to disunite the elements of bodies, as a test or measure of the intensity of chemical union. By the use of the multiplier, it would now be easy to apply this test; and accurate researches on the connexion of what may be called the electro-dynamic relation of bodies to their combining masses or proportional numbers, will be the first step towards fixing chemistry on the permanent foundations of the mathematical sciences.'

On 28 September in discussing the results of experiments in which a current is passed through a sulphuric acid voltameter, and various solutions in series with it, Faraday decided that in aqueous solution the current is probably carried by hydrogen and oxygen, these being always the primary products of electrolysis.

When, therefore, metallic solutions are decomposed the metals are evolved not by the current of electricity but by the hydrogen evolved at the N. Pole.... Hence it will probably follow that in these cases the metal is an equivalent of the hydrogen because it is produced chemically by the hydrogen, and therefore such effects *will not PROVE* the equivalent character of the products of true electro-chemical decomposition.... Perhaps fused nitre will be a good salt to compare by current with decomposition of water. Or fused chloride of lead or tin.

These experiments with fused salts, which removed any doubt as to whether the elements liberated during electrolysis were primary or secondary products, were not carried out until December, but doubtless in the interval Faraday was making plans for them.

On 17 December he wrote in his note-book: 'Preceeded to decompose dry chlorides, oxides, etc., to ascertain if there also the decomposition was definite and what the equivalent numbers would be.' So quickly was the final stage in the investigation accomplished that on 9 January 1834, the paper containing the Laws of Electrolysis was communicated to the Royal Society. In the first experiment on 17 December fused stannous chloride in a glass tube was decomposed with platinum wire poles with a voltameter in series, and the weight of the tin liberated compared with the weight of water (0·26486 gr) decomposed in the voltameter. '1·76 of tin had been electrochemically evolved at the exode, and of course a corresponding portion of chlorine at the cisode. Now

$$\underset{W}{0\cdot 26486} : \underset{T}{1\cdot 76} :: 9 : 69\cdot 805 \text{ the tin.}$$

'The number for tin is given 58 which is very near indeed for a first experiment, and shows that the electro-chemical equivalent is the same as the chemical equivalent here.' Note Faraday's first efforts at a new terminology, 'exode' and 'cisode'. Later on the same day the word 'pole' which suggests the idea of attraction or repulsion, was struck out and 'electrode' written above it for the first time.

On the following day another experiment with stannous chloride gave a value of 53·833 for tin. Two similar experiments with fused lead chloride gave electrochemical equivalents for lead of 105·11 and 97·22, the chemical equivalent being 103·5 or 104. Faraday wrote of the lower value: 'Hence it is too little, but still so near as to establish the principle of electrochem equivalents.'

On 19 December no experiments are recorded, but doubtless Sergeant Anderson was helping Faraday to set up apparatus, and fresh glass vessels with fused-in electrodes had to be blown for each experiment. A few extracts from the notebook on that day show the activity and range of Faraday's mind:

1192. With regard to *INTENSITY* and its meaning, etc., Define intensity if possible and state its relation to *quantity, time and conducting power*.

1195–1200. Nervous agency of Electricity.

Michael Faraday

1207. In the table I mean Real Electro chemical equivalents not hypothetical for we shall else outrun fact and lose the information directly before us. . . . I must keep my researches really *Experimental* and not let them deserve anywhere the character of *hypothetical imaginations*.
1212. Search for Fluorine by using plumbago Pos. Pole acting on a fluoride.
1213. This process may finally give rise to some very good processes of analysis in determining weights or at least to some excellent modes of comparing weights of metals . . . *a good principle of analysis* for it will hold probably in salts as well if properly selected and may use mercury electrodes when convenient.

A remarkable anticipation of modern methods of electrolytic analysis.

With the exception of Christmas Day, determinations were made every day. On 26 December Faraday comments in his diary on the enormous quantity of electricity required to decompose a small amount of water. Later he estimates that 800 000 charges of a Leyden battery, each one of which would suffice to kill a cat, 'would be necessary to supply electricity sufficient to decompose a single grain of water; or, if I am right, to equal the quantity of electricity which is naturally associated with that grain of water, endowing them with their mutual chemical affinity!'

The experiments on the electrolysis of fused salts were difficult and often gave inconclusive results, but confirmatory evidence was obtained from lead borate and iodide.

Faraday was anxious to extend the work to the deposition of metals from aqueous solutions, and he found that zinc deposited on a platinum electrode gave an electrochemical equivalent of 34·08 while the loss in weight of amalgamated zinc in a platinum-zinc cell compared with the weight of hydrogen evolved gave in two experiments equivalents of 30·2 and 32·31. 'Excellent', he writes after the latter result, to which he attached great importance, as it was the first occasion on which he had compared the amount of chemical action in a voltaic cell with that produced by the current in the external circuit. This confirmed his conviction

that the quantity of electricity which, being naturally associated with the particles of matter, gives them their combining power, is able, when thrown into a current, to separate those particles from their state of combination; or, in other words, that the electricity which decomposes, and that which is evolved by the decomposition of a certain amount of matter, are alike.

The harmony which this theory of the definite evolution and the equivalent definite action of electricity introduces into the associated theories of definite proportions and electro-chemical affinity is very great. According to it, the equivalent weights of bodies are simply those quantities of them which contain equal quantities of electricity . . . it being the ELECTRICITY which determines the equivalent number, because it determines the combining force. Or if we adopt the atomic theory or phraseology then the atoms of bodies . . . have equal quantities of electricity naturally associated with them. But I must confess I am jealous of the term atom, for though it is very easy to talk of atoms, it is very difficult to form a clear idea of their nature.

On this occasion Faraday was evidently so excited at the success of these experiments that he anticipated his maxim of 'Work, Finish, Publish' by communicating his results to the Royal Society before the investigation was completed. We know from his diary that his experiments showing that the amount of chemical action in a voltaic cell is equivalent to the hydrogen liberated by the voltaic current were not completed until 10 January. However the paper in the *Philosophic Transactions* is dated at the end, 'Royal Institution, December, 31st, 1833', and it was received on 9 January 1834 and read 23 January, 6 and 13 February. We know also from Professor Sydney Ross's paper, 'Faraday consults the Scholars', in *Notes and Records* for 1961, that Faraday had not agreed with Whewell the new nomenclature he used, in the first part of the paper in the *Philosophic Transactions*, until 9 May 1834. So clearly the paper must have undergone considerable revision and extension after the date on which it was originally submitted. This is evidence of his mental excitement at the results he had obtained, and there is an emotional quality running through the pages of this classic paper, which was the culmination of three years of continuous study.

This paper is the most important of Faraday's contributions to electrochemistry, and in it he summarizes all his previous work. He begins by introducing the new terminology which he

Michael Faraday

devised with the help of Whewell for the sake of greater precision of expression, and all his new names—electrode, anode, cathode, ion, anion, cation, electrolyte, and electrolysis—we use today with the significance which Faraday gave to them. After a short account of the conditions necessary for electrochemical decomposition, he describes his new volta-electrometer and the evidence that led him to the conclusion that the amount of chemical action is dependent solely on the amount of electricity that passes through it. He next discusses whether the products of electrolysis are primary or secondary, and gives his evidence for the identity of chemical and electro-chemical equivalents.

The researches had strengthened enormously the evidence for his First Law of Electrolysis—'The Chemical power of a current of electricity is in direct proportion to the absolute quantity of electricity which passes'; and they had established the Second Law—'Electro-chemical equivalents coincide and are the same with ordinary chemical equivalents.' The exactness of these two laws has been confirmed by every subsequent investigation.

Faraday then spent a month carrying out a large number of experiments on the intensity required to produce electrolysis of different solutions by varying the number of cells in the battery and seeing how many were required to electrolyse various compounds in solution or in a fused state. One of his difficulties was that he had no unit to measure by, and on 10 February he notes, 'The power of decomposing water a good *unit* of intensity in voltaic apparatus'.

The general result of these experiments was to strengthen Faraday's conviction that the source of the current was the chemical reaction taking place in the voltaic cell and not the mere contact of two metals, and indeed he proved that a current is produced without such contact by interposing a slip of paper moistened with potassium iodide solution between the metals. His thoughts were concentrated on the relation between chemical action and the production of electricity, and he realized that whether a current passes or not depends on the relative magnitudes of the chemical affinities of the reactions taking place in the battery and in the electrolytic cells. In discussing this problem on 12 February he writes:

The whole arrangement seems beautifully to show that antagonism of the *chemical powers* and the Electromotive parts with the *chemical powers* and the interposed parts. The first are producing electric effects, the second opposing electric effects, and the two seem equipoised as in a balance, and in both cause and effect appear to be identical with each other. Hence chemical action merely electrical action and Electric action merely chemical.

Again on 19 February:

Affinity is action at both points, but as is it were connected or related by the current of electricity in the communicating wires, or in other words affinity is electricity and vice versa.

Three days later he wrote:

We seem to have the power of deciding in certain cases of chemical affinity [as of zinc with the oxygen of water] which of two modes of action of the *one power* shall be exerted. In the one mode we can transfer the power on, it being able to produce elsewhere its equivalent of action; in the other it is not transferred on but exerted at the spot. The first is the case of Voltaic Electric production, the other the ordinary cases of chemical affinity. But both are chemical actions and due to one power or principle.

In other words, Faraday saw that a chemical reaction can be carried out in two ways, either by means of a voltaic cell in which the reactants are separated by an electrolyte, or by their direct contact, and further, he identified the electromotive force of the cell with the chemical affinity of the reaction. Half a century was to elapse before the conception of chemical affinity assumed a definite form in chemists' minds, but here Faraday anticipates our modern interpretation. His method of reasoning, too, is an instinctive recognition of the Law of Conservation of Energy, and it was in connection with the chemical theory of the cell that he wrote in 1840: 'In no case ... is there a pure creation or production of power without a corresponding exhaustion of something to supply it.'

All these results were collected in a paper 'On the Electricity of the Voltaic Pile', and communicated to the Royal Society on 7 April. This was really the last of Faraday's great contributions to electrochemistry, although a few papers of minor importance came later.

Faraday's researches had a most profound and immediate

effect on the progress of electrochemistry. Within two years he changed the whole aspect of the subject and gave it a coherent structure and a quantitative basis, as a result of his laws of electrolysis, his new ideas which were crystallized in his new nomenclature, and his association of the intensity of the voltaic cell with the chemical affinity of the reaction taking place in it.

It is remarkable, too, how many of Faraday's ideas and discoveries had a decisive influence on the development of electrochemistry in the nineteenth century, and have even today a direct bearing on modern theories. His experiment on the transfer of sulphuric acid during electrolysis inspired the investigation of Daniell, which proved that the current in aqueous solution is carried by the ions of the solute, and not as Faraday supposed by the ions of hydrogen and oxygen. Later its development by Hittorf led to the conception of transport numbers which has played so important a part in the theory of solutions.

Again in 1834, Faraday pointed out that if we adopt the atomic theory or phraseology then 'the atoms of bodies... have equal quantities of electricity associated with them'. Had he been a believer in the atomic theory he might have made the deduction that electricity, like matter, is atomic in nature. It was left most appropriately to Helmholtz in his Faraday lecture in 1881, to point out this most startling result of Faraday's laws: 'If we accept the hypothesis that the elementary substances are composed of atoms, we cannot avoid concluding that electricity also, positive as well as negative, is divided into elementary portions, which behave like atoms of electricity.' Thus Faraday's laws led directly to the conception of the electron.

One further example—Faraday first pointed out the enormous size of the electrical charge carried by each ion. Chemists and physicists lost sight of this fact until Helmholtz recalled their attention to it fifty years later in his Faraday lecture, and showed that the attractive force between the electrical charges associated with hydrogen and oxygen is 71 000 billion times greater than the gravitational attraction between their masses. Again many years passed before first Milner and then Debye and Hückel showed that the magnitude of these interionic forces could account for the so-called anomalies of strong electrolytes. There is thus a direct link between Faraday and the most modern theory of solutions.

After 1834 Faraday's researches were mainly physical but each in turn has had its influence on physical chemistry, an influence that is still felt today. In 1837 came his paper on Induction, dependent on the property of the dielectric medium which he defined as its Specific Inductive Capacity. Faraday thought of the particles of the dielectric as polarized, which finds its modern counterpart in their dipole moments.

In September 1845, at long last, he found the effect of magnetism on light in the rotation of the plane of polarization in a magnetic field, one of his most exciting discoveries. And although as a so-called additive property magnetic rotation did not fulfil its early promise, today it has its value in determining paramagnetic susceptibilities connected with the spin of electrons.

The discovery of diamagnetism in November 1845 led up to his paper on the 'Magnetic Condition of all matter'. This opened a new line of investigation which today finds many applications in the study of the detailed structure of atoms and molecules.

In 1846 came Faraday's boldest speculation in his paper on 'Thoughts on Ray-Vibration', foreshadowing the electromagnetic theory of light. Clerk Maxwell's mathematical interpretation of Faraday's physical conceptions revealed, as von Helmholtz said, the quite wonderful sincerity and intellectual precision with which Faraday performed in his brain the work of a great mathematician without using a single mathematical formula.

With the marvellous intuition that guided Faraday to the fundamentals of any problem and with his genius for experiment, it is no wonder that physical chemistry in so many fields still bears the impress of his mind. So often Faraday's experiments were not the exploration of an uncharted field, but the verification for himself of his instinctive recognition of Nature's ways. As Kohlrausch said of him 'Er riecht die Wahrheit'—he smells the truth; or in Tyndall's words, 'Faraday was more than a philosopher; he was a prophet.'

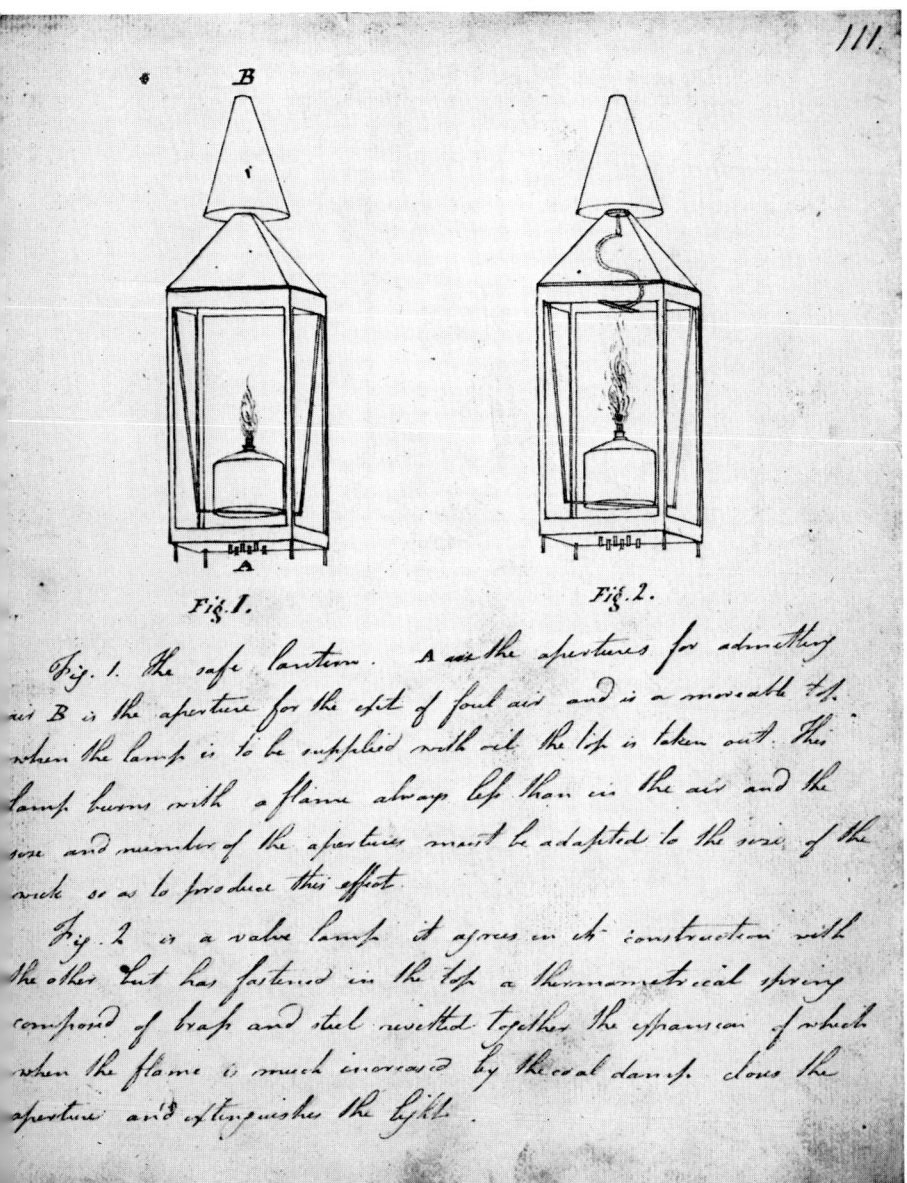

Faraday's drawing of the safety-lamp

Jöns Jakob Berzelius
From the portrait by A. J. Way

MICHAEL FARADAY
From the portrait by A. Blaikley

SUNTO DI UN CORSO
DI
FILOSOFIA CHIMICA
FATTO

Nella R. Università di Genova

DAL PROF. S. CANNIZZARO

NOTA

SULLE CONDENSAZIONI DI VAPORE

DELL'AUTORE STESSO

PISA
TIPOGRAFIA PIERACCINI
1858

Title-page of Cannizarro's classic pamphlet

7

FARADAY'S SUCCESSORS AND THE THEORY OF ELECTROLYTIC DISSOCIATION†

During the century that has elapsed since Faraday's discoveries, the conduction of electricity in solutions has remained one of the central problems of chemistry and physics, and I propose now to trace briefly the steps by which we have arrived at our present position and to recall the chief landmarks in the history of the ionic theory. There have been three main phases in the development of the problem. First, the discovery of the general relationships between the conductivity of solutions and the concentration and nature of the dissolved substances, due mainly to the work of Hittorf and Kohlrausch; secondly, the recognition of the relations of these facts to the general theories of chemistry in the classical ionic theory of Arrhenius; and, lastly, the quantitative explanation of the properties of electrolytes in the mathematical theory due to Milner, and to Debye and Hückel.

The direct successor to Faraday was Daniell, who studied in detail the changes in the concentration of electrolytes at the electrodes produced by electrolysis, which had first been observed by Faraday. His papers were published in three letters to Faraday in 1840–4, and their most important result was to show that the current in aqueous solutions is actually carried by the ions of the solute and not, as Faraday had supposed, by the ions of hydrogen and oxygen. Daniell failed to find any reasonable explanation of his transference data.

Ten years later this aspect of electrolysis was the subject of a classical investigation by Hittorf on 'The Migration of Ions during Electrolysis', which was of the great significance for the

† From the Presidential Address to Section B of the British Association in 1931.

theory of solutions. Hittorf realized that the changes in concentration round the electrodes could only be explained on the assumption that the ions move with different velocities, and he showed how their relative speeds could be calculated from the change in concentration, a very remarkable achievement in 1853. Not only did Hittorf show great theoretical acumen, but he was an outstanding experimentalist. He devised the methods by which transport numbers are still determined, and so accurate and comprehensive were his results, that until recently they were the main source of our knowledge of transport numbers. Hittorf, like Faraday, thought that the transference of the two ions through a solution was by means of a Grotthus chain, a view which now seems to us difficult to reconcile with the fact that the ions move at different speeds.

Simultaneously, with the work on transference, a number of observers like Wheatstone, Wiedemann, and Beetz, were studying the conductivity of solutions in the light of Ohm's Law. Ohm's papers were actually published before Faraday's, but Faraday knew no German and we are left to speculate as to what would have been the effect on him of realizing the mathematical relationship between the factors that were so often in his mind. The problem was complicated by the polarization of the electrodes, and progress was slow until Kohlrausch solved this difficulty by the use of alternating current.

Kohlrausch must always remain the outstanding figure in the experimental study of the conductivity of solutions. For forty years his genius for exact measurement, his fine critical brain, and his untiring industry were devoted mainly to this work. He devised the experimental methods we use to-day, and he surveyed for us with unerring accuracy the field of aqueous solutions. But his work went far beyond the mere collection of data. In 1876 he recognized the Law of the Independent Mobility of Ions. In 1878 he introduced that most convenient term, the equivalent conductivity of a solution, and he established the modern conception of ionic motion by calculating the mean velocities of the ions and showing that they were of the magnitude that would be expected if the ions were particles of molecular dimensions moving in accordance with the laws of hydrodynamics. In 1886, Lodge confirmed his calculation by measuring the actual velocities of the ions in a known field.

Faraday's Successors

Kohlrausch quickly realized the theoretical importance of work on dilute solutions, devising special methods for their study, and perhaps it is only those who have tried to repeat his measurements, both in this field and on the conductivity of pure water, who can really appreciate the experimental skill which attained such accuracy before the days of thermostats. To those of us who work in this field his papers are a constant source of inspiration, and may I acknowledge here my personal debt to him for the encouragement and the ready help which he gave so generously to an unknown beginner?

The evidence of Kohlrausch and Hittorf for the independent movement of the ions in dilute solutions was so clear that to us it seems surprising that the ionic theory as we know it today was not immediately forthcoming. As early as 1857 Clausius had pointed out that, since the resistance of a solution obeys Ohm's Law, no work can be done by the current in separating the molecules into ions, which must therefore be already in existence, as a result, he supposed, of collisions between molecules. But just as the *a priori* conceptions of chemists had prevented the acceptance of Avogadro's hypothesis for nearly half a century, until the chemical evidence in its favour was overwhelming, so again convergent evidence from different quarters, chemical as well as physical, was necessary before the idea that electrolytes might be largely dissociated into electrically charged ions was even entertained. The difficulty in chemists' minds was twofold—first, how could mere solution separate a molecule into the ions of elements with a great affinity for one another; and secondly, how could the ions remain in presence of water without chemical action?

The next phase is almost too well known to need recalling. In 1883 Arrhenius, working on dilute aqueous solutions, concluded that the equivalent conductivity increases with dilution because the proportion of conducting molecules increases. He then found a correspondence between the conductivities of acids and their strength as determined thermochemically by Berthelot, or by the method of displacement. This suggested to him that the molecules which are active as regards conductivity are active chemically, and it occurred to him that these active molecules are dissociated, since this would explain why the heats of neutralization of all strong acids and bases are the same. He communicated this idea to Ostwald, who was then working on the catalytic activity of different acids, and they found that the catalytic activities of these acids was roughly proportional to their conductivities. In 1886 van 't Hoff published his memoir on the analogy between dilute solutions and gases, in which he drew attention to Raoult's measurements of the freezing-points of aqueous solutions, which showed that the influence of one molecule of an electrolyte like potassium chloride was double that of a molecule of a non-electrolyte such as alcohol. Arrhenius writes:

> it was quite clear to me that I might dare to say that all those substances which are active, that is all electrolytes, consist of two molecules and not of one; that is, sodium chloride is composed of two molecules, the sodium ion and the chlorine ion. Then the theory of electrolytic dissociation was expressed without any restriction [1887]. I had then a threefold basis for my conclusion, the chemical one and the electrical one, and then the thermodynamical one, regarding the freezing-point. On a foundation of three points you may construct a very solid building.

It is interesting that both Planck and van 't Hoff put to Arrhenius the difficulty that if a salt is partially dissociated into ions, then the equilibrium between the ions and molecules should be in agreement with Guldberg and Waage's Law of Mass Action, while they found by calculating the degree of dissociation by Arrhenius's formula $\alpha = \Lambda_c/\Lambda_0$ that this was not the case. Arrhenius suggested that a better test would be made by considering the dissociation of weak acids which varied over a much wider range and the results, as we know, confirmed his theory, and were embodied in Ostwald's Dilution Law.

On its publication in 1887 the ionic theory met with violent opposition, but, in the hands of Arrhenius, Ostwald, van 't Hoff, and Nernst, its value was quickly established in the most varied fields: the theory of concentration cells and of liquid junction potentials, the behaviour of weak acids and bases and the hydrolysis of salts, the properties of indicators, the dissociation of water, and the theory of qualitative and quantitative analysis. Entirely fresh light was thrown on all these problems, and in many instances they admitted of quantitative explanation for the first time. It is true that the reason for ionization and for the stability of ions remained unknown for many years, until the discoveries of Rutherford and Moseley had made possible Bohr's application of the quantum theory to the problem of atomic structure. Then the connection between electricity and chemical affinity, which chemists had been seeking since the days of Berzelius and Faraday, suddenly became clear, and we learnt the cause of ionization and of the stability of ions in solution. Metallic sodium reacts with water in order to give up an electron; the sodium ion having already lost an electron is no longer reactive.

But in spite of the many triumphs of the ionic theory, and its success so far as weak electrolytes are concerned, the discrepancy between the behaviour of strong electrolytes and the mass law, pointed out by Planck and van 't Hoff to Arrhenius, still remained. Kohlrausch had shown that in very dilute solutions the relationship between the equivalent conductivity and concentration of salts was expressed by the equation

$$\Lambda_c = \Lambda_0 - kc^{\frac{1}{2}} \quad (k = \text{constant})$$

which is incompatible with the mass law. At the beginning of this century the outstanding problem of the ionic theory was this so-called anomaly of strong electrolytes. Up to this point the whole development of the ionic theory had come from the experimental side, and every advance had originated in some new discovery. But the solution of the final problem came not from experiment but from the mathematical physicists, who thus repaid to chemistry the debt which physics owed to Faraday.

Faraday had often drawn attention to 'the enormous electric power of each particle or atom of matter,' i.e. the large size of the ionic charge. Helmholtz in 1881, in his Faraday lecture, made a calculation showing that the attractive force between the electrical charges associated with hydrogen and oxygen is 71 000 billion times greater than the gravitational attraction between their masses. It would seem obvious that forces such as these must affect the behaviour of ions, and from time to time suggestions came from chemists—van Laar in 1900, Bjerrum in 1906, Sutherland in 1907—that strong electrolytes were completely dissociated, and that the variations of equivalent conductivity with concentration were due not to a change in the degree of dissociation, but to the varying effect of the interionic forces. Bjerrum had found that the molecular colour of chromium salts was independent of dilution in the absence of complex ions, and explained this on the basis of complete dissociation.

One of the earliest supporters of the ionic theory was Nernst, who made very substantial contributions to it in his theory of concentration cells and of diffusion potentials. Milner, working in his laboratory at Göttingen in 1898, was attracted by the problem of strong electrolytes, and attacked it from the point of view of the interionic forces. The mathematical difficulties were great, and it was not until 1909 that they had been overcome sufficiently to admit of the calculation of the change of total internal energy with dilution. Milner was the first to realize that the ions cannot be distributed at random in a solution since, owing to the Coulomb forces, there must be an excess of positive ions in the neighbourhood of a negative ion and vice versa. Thus each ion can be considered as surrounded by a spherical ionic atmosphere, the density of which decreases with the distance from the ion. The electrical potential at the surface of the central ion will therefore be affected by the ionic atmosphere, and by taking into account the changes of potential with dilution, Milner was able to calculate the freezing-point depression of an electrolyte at different dilutions, assuming that it was completely dissociated. He did not, however, attack the problem of conductivity.

Another ten years elapsed without further progress until a discussion at Zurich in 1921 of the ingenious, but erroneous, theory devised by Ghosh attracted the attention of Debye to a field that was entirely new to him. That lucky accident tempts us to speculate on the potential value of a doubtful hypothesis, for how would the problem stand now if Ghosh's papers had not seen the light of day?

Debye, who at the time did not know of Milner's work, found a simple mathematical solution of the problem by applying Poisson's equation to the relation between the average potential and the density of charge at any point in the sphere surrounding the central ion, and he showed that the distribution of the charge in the ionic atmosphere depends on the square root of the concentration; thus, explaining why such diverse properties of solutions as activity coefficients and equivalent conductivities are functions of $c^{\frac{1}{2}}$. With the collaboration of Hückel, the whole

Faraday's Successors

problem was attacked in detail, and in 1923 they succeeded in calculating the effect of the ionic atmosphere on the mobility of the ion.

The conductivity of a solution depends on the number of ions that it contains and on their mobility. The classical theory of Arrhenius considers the effect of ionic dissociation on the former factor, whilst the Debye–Hückel theory considers the effect of the interionic forces on the latter.

When an ion moves under an external potential gradient, it has to build up continuously a fresh atmosphere in front of it while the atmosphere behind it has to die away, but since the ionic atmosphere takes a finite time to form or disperse, there will always be an excess of ions of the opposite sign in its rear, and consequently it will be subject to a retardation due to the dissymmetry of the atmosphere, which depends on the velocity with which it is moving. Further, as ions of opposite signs are moving in opposite directions and as both carry with them a certain amount of solvent, the viscous resistance to the motion of the ions will be greater than if the solvent were at rest. Thus, both these effects reduce the mobility of the ion below its value at infinite dilution, and Debye and Hückel arrived at the following equation for the variation of the equivalent conductivity of a z-valent binary electrolyte in a solvent of dielectric constant D at a temperature T,

$$\frac{\Lambda_0 - \Lambda_c}{\Lambda_0} = \left\{ \frac{K_1}{(DT)^{3/2}} w_1 + \frac{K_2}{(DT)^{1/2}} w_2 b \right\} \sqrt{2c} \qquad (1)$$

in which the first and second terms on the right-hand side are the dissymmetry and viscosity terms respectively, K_1 and K_2 are universal constants, w_1 and w_2 are valency factors, and b is the average radius of the ions. This reduces to

$$\Lambda_c = \Lambda_0 - x\sqrt{c} \qquad (2)$$

which is identical in form with the empirical equation of Kohlrausch. Comparison with experimental results shows that the coefficient of $c^{\frac{1}{2}}$ in equation (1) is of the right order of magnitude if a value is assumed for the ionic radii of 10^{-8} cm in accordance with X-ray data. But if the value of b is calculated from the ionic mobilities at infinite dilution, assuming Stokes's law to hold, the observed and calculated coefficients are not in exact agreement, e.g. for potassium chloride solutions in water at 25° they are 0·461 and 0·547 respectively. This difference is greater than the experimental error.

Onsager, in 1926, pointed out that Debye and Hückel, in calculating the effect of the dissymmetry of the atmosphere around a moving ion, had neglected the Brownian movement of the ion and that a correcting factor of $2-\sqrt{2}$ must be introduced on this account. His final equation has the same general form as Debye and Hückel's, and when numerical values are inserted for the universal constants, it becomes for a z-valent binary electrolyte

$$\Lambda_c = \Lambda_0 - \left[\frac{0.986 \times 10^6}{(DT)^{3/2}}(2-\sqrt{2})\, z^2 \Lambda_0 + \frac{58 \cdot 0}{(DT)^{1/2}\eta}\, z\right]\sqrt{2zc} \quad (3)$$

where η is the viscosity of the solvent.

For various solvents at 25° this equation becomes for uni-univalent electrolytes:

Water: $\Lambda_c = \Lambda_0 - (0 \cdot 228\, \Lambda_0 + 59 \cdot 8)\sqrt{c}$
Methyl alcohol: $\Lambda_c = \Lambda_0 - (0 \cdot 957\, \Lambda_0 + 158 \cdot 1)\sqrt{c}$
Ethyl alcohol: $\Lambda_c = \Lambda_0 - (1 \cdot 256\, \Lambda_0 + 87 \cdot 8)\sqrt{c}$

For sodium chloride in each solvent these equations become:
Water: $\Lambda_c = 126 \cdot 4 - (28 \cdot 8 + 59 \cdot 8)\sqrt{c}$
Methyl alcohol: $\Lambda_c = 97 \cdot 0 - (93 + 158 \cdot 1)\sqrt{c}$
Ethyl alcohol: $\Lambda_c = 43 \cdot 0 - (54 + 87 \cdot 8)\sqrt{c}$

These equations show that the dissymmetry term and the viscosity term (the first and second coefficients of $\sqrt{c}$ respectively) are of the same order. Their relative magnitude varies, however, with the properties of the solvent and with the ionic velocities of the ions present.

The fundamental idea of the new theory, the existence of an ionic atmosphere with a finite time of formation and dispersion, has now been definitely established by the work of Wien on conductivities at high electromotive forces, and of Debye, Falkenhagen, and Sack on conductivities at high frequencies. With a sufficiently great ionic velocity the atmosphere would not have time to form, while with a high enough frequency its dissymmetry would vanish owing to the negligible displacement of the ion. In both cases the experimental results showed a satisfactory agreement with theory.

The Debye–Hückel–Onsager equation enables us to calculate the change in equivalent conductivity with dilution for any electrolyte in any solvent provided that we know the ionic mobilities and valencies involved and certain physical constants of the solvent, but in comparing the theory with the results of conductivity determinations, the assumptions underlying it must be remembered, viz.:

1. That the electrolyte is completely dissociated into point ions and that all interionic forces except Coulomb forces can be neglected.
2. That corrections for the overlapping of ionic atmospheres can be neglected.
3. That the solvent between the ions retains the properties of the pure solvent.

These conditions can only be fulfilled in dilute solutions, and a great deal of work has been done recently with uni-univalent electrolytes in a number of solvents to test the theory in the dilute range.

The results show that there is a close approach to a linear relation between Λ_c and $c^{\frac{1}{2}}$, as required by the theory, for strong electrolytes in solvents with a dielectric constant greater than 20. In water, methyl and ethyl alcohols, nitromethane, and acetonitrile the slopes of the conductivity curves agree well with theory for a number of electrolytes, and any large deviations are such that they can be explained by ionic association. In fact the body of evidence now available seems sufficient to justify the use of the Debye–Hückel–Onsager equation to represent the behaviour of a perfect electrolyte in dilute solution and we have, therefore, a new means of judging whether an electrolyte is appreciably associated or not.

The term ionic association marks the contrast between the old outlook and the new. Arrhenius thought of the act of solution as separating the molecules into ions. We know now that most salts are already ionized in the crystalline state, and the question that concerns us is whether the condition of complete ionization persists on solution of the crystal. In many cases the conductivity of an electrolyte is less than we should expect from the Debye–Hückel equation, indicating that some modification of the state of configuration of the ions has occurred which involves a decrease in the conductivity. This may be due either to the formation of a covalent linkage between the ions or to a modification of their distribution leading in the extreme case to the formation of an ion pair as suggested by Bjerrum. The term ionic association is used to cover both possibilities.

The influence of the solvent on the properties of an electrolyte is illustrated very clearly by a comparison of the behaviour of uni-univalent salts in water and in non-aqueous solvents. In water they are all strong electrolytes with a surprisingly uniform behaviour, as shown in Kohlrausch's classic diagram in the *Zeitschrift für Electrochemie* for 1907. In non-aqueous solvents, however, Walden's comprehensive investigations have shown that individual differences begin to appear as the interionic forces increase and the specific affinities of the ions are brought to light. The question naturally arises as to whether the extent of the ionic association is determined entirely by the interionic forces, i.e. by the dielectric constant of the solvent. This is clearly not the case, since a number of salts are strong electrolytes in methyl alcohol and weak electrolytes in nitromethane, which has a higher dielectric constant (37 as against 30·3). Walden and Ulich have pointed out that, in general, non-hydroxylic solvents such as the nitro-compounds and acetone accentuate the individual differences between electrolytes, while the hydroxylic solvents suppress them. The reason for this is probably to be found in a difference in the nature of the solvation of the ions in the two classes of solvents. There is abundant evidence that the ions in solution have a number of solvent molecules attached to them either by co-ordinate linkages or as a result of the dipole character of the solvent, and these solvent atmospheres exert an important influence on the behaviour of the ions. For instance, the evidence of the ionic mobilities of the alkali metals and of Washburn's transference experiments leaves no doubt that the effective size of the ion is determined by its diameter, including any envelope of solvent which it carries with it.

Bjerrum has considered the effect of ionic size on the probability of the formation of ion pairs, which would contribute nothing to the conductivity of a solution, and has shown that if the sum of the radii of the two ions is below a certain value, the number of ion pairs will increase rapidly. Hence, the solvation of ions may have an important effect in preventing the ions from coming near enough together to form ion pairs.

Sidgwick has considered solvation from the electronic standpoint, and has emphasized the importance of the donor and acceptor properties of hydroxylic solvents in enabling them to form co-ordinate links with both anions and cations. Non-hydroxylic solvents, however, like nitromethane and acetone, can only form co-ordinate links with cations, leaving the anions chemically unsolvated, and it is very significant that in such solvents lithium salts are weak electrolytes, while in hydroxylic solvents they are less associated than the salts of the other alkali metals. It would thus appear that the existence of a chemical link between the solvent molecules and both ions is of cardinal importance in preventing ionic association. This view of the protective action of hydroxylic solvent molecules is confirmed by the effects of small quantities of water on the conductivity of solutions in nitromethane and the other non-hydroxylic solvents. For example, the conductivity of lithium thiocyanate, a weak salt, is increased 60 per cent by the addition of 0·1 per cent of water, while that of tetra-ethylammonium iodide, a strong electrolyte, is only increased 0·22 per cent by a similar addition.

Until recently, most of our ideas about electrolytic solutions were based on experience in water, and this gave quite a false impression of the simplicity of the problem, since in water, thanks to its high dielectric constant and to the protection afforded by the chemical solvation of both ions, all uni-univalent salts exhibit almost ideal behaviour. But a survey of non-aqueous solutions reveals at once a much more complex situation, in which the chemical nature of the solvent and the affinities of the ions are often the predominant factors. Thus, a purely physical theory (*pace* Faraday), like that of Debye and Hückel, while invaluable in explaining and predicting the behaviour of an ideal electrolyte, is far from giving a complete picture of electrolytic solutions even in the dilute range, since it leaves out of account the chemical nature both of the ions and of the solvent molecules.

Looking back over the century we see how the mechanism of electrolytic conduction has gradually been disclosed to us, and how in recent years the many-sided influence of the solvent has come more and more into prominence. By its dielectric constant the solvent determines the magnitude of the interionic forces. By its power of solvating the ions it exerts a decisive influence on the extent to which they form ion pairs or undissociated molecules. And lastly by its viscosity, as well as by the extent of solvation, it determines the mobilities of the ions. The task of the immediate future is to discover the precise nature and extent of the solvent atmosphere around the ions which exerts such an important influence on their properties. And so today we find ourselves face to face with a new phase of the problem, and we can repeat with Faraday the words near the close of his great paper on electrolysis: 'Indeed, it is the great beauty of our science, CHEMISTRY, that advancement in it, whether in a degree great or small, instead of exhausting the subjects of research, opens the doors to further and more abundant knowledge, overflowing with beauty and utility, to those who will be at the easy personal pains of undertaking its experimental investigation.'

8

STANISLAO CANNIZZARO, F.R.S. (1836-1910) AND THE FIRST INTERNATIONAL CHEMICAL CONFERENCE AT KARLSRUHE IN 1860†

My interest in the Karlsruhe Conference dates from the day in 1899 when I bought for one mark the copy of Cannizzaro's classic pamphlet, *Sunto di un corso di filosophia chimica*, which Pavesi gave to Hermann Kopp after the last session of the Conference, containing some notes in Kopp's handwriting.[1] I have always been puzzled as to why the Conference, in spite of Cannizzaro's presence and his eloquent speeches, failed to achieve its object, and yet when they read his pamphlet on the way home, as Lothar Meyer says, 'the scales fell from my eyes'.[2] At long last I think I see how it happened and on whose shoulders the responsibility rests.

During the years 1825-60 the efforts of most of the outstanding chemists—Liebig, Wöhler, Dumas, Laurent, Gerhardt, Wurtz, Williamson, Kekulé, Kolbe, and Frankland were directed to the organic field, the compounds of carbon, where they found a rich harvest. Each of them had his own theory which was of special value to him in stimulating his line of attack. As a result, rapid progress was made in a number of different directions, which were bound to lead eventually to a common understanding of the structure of carbon compounds. Probably at no time in the history of chemistry was there such bitter personal animosity between these holders of rival theories, culminating in the unfortunate persecution of Laurent and Gerhardt and the death of Laurent from sheer want. So great

† *Notes and Records of the Royal Society of London*, **21**, p. 56 (1966).

was the lack of a common point of view that in 1859 Kekulé filled a whole page of his textbook with different formulae of such a simple compound as acetic acid and the formula of water was written in four different ways, H_2O, HO, $\overline{HO}$, H_2O_2.

It was this confusion of ideas that stimulated the young Cannizzaro in 1858 to give a course of lectures to his students in which he tried to rationalize the conceptions of atom and molecule and of atomic and molecular weight. A year later Kekulé had the idea of initiating an international chemical conference to try to reach some agreement in order to end the chaos that was bedevilling progress.

Cannizzaro was born in Palermo in 1826 and he inherited from his parents much of the fiery temperament of the Sicilians. He first studied medicine at the University of Pisa and later changed to chemistry, in 1850 becoming Piria's assistant. In his researches on benzyl alcohol he discovered the reaction which bears his name. From 1853 to 1858 he held a professorship at the Technical Institute of Allessandra and in 1858 was appointed to the Chair of Chemistry in the University of Genoa. It was there that he gave the classic course of lectures on atomic and molecular theory that is a landmark in the history of chemistry. Cannizzaro saw so clearly that this confused state of chemical theory was due to the reluctance of chemists to accept wholeheartedly the logical conclusions from the work of Gay-Lussac and Avogadro owing to their preconceived ideas on one aspect or another. He was anxious that his students should be saved from the confused thinking of the period. Fortunately we know the run of Cannizzaro's mind from the précis of the lectures sent to de Luca which was printed in the *Nuovo Cimento* in May 1858. The lectures were based on a historical exposition of the course of events, starting with the work of Gay-Lussac, Avogadro, and Ampère, making logical deductions from the known experimental facts. Cannizzaro next showed how Berzelius, misled by his own electrochemical theory and dualistic approach, disregarded the distinction made by Avogadro between the atoms and molecules of elementary gases and continued to regard their elementary particles as monatomic. Basing his approach solely on molecular weights determined by vapour densities and on the proportions of the elements present in the molecular weights of elements and compounds, Cannizzaro showed the conclusive nature of the criteria regarding the atomic weights of the elements which agreed with the values assigned to them by Berzelius in 1826 with the exception of the alkali metals and silver.

Cannizzaro next showed that his values for atomic weights were confirmed by the values of the specific heats of elements and compounds following Dulong and Petit and Regnault's later work.

Turning then to organic chemistry and following Gerhardt and Laurent he showed how the vapour densities of organic compounds were all consistent with his atomic weights and with Gerhardt's formulae. He pointed out that Gerhardt's assumption, that metallic monoxides all had the same formulae as water, had led him to assign wrong values to the atomic weights of many metals. The formulae and the equations with which Cannizzaro illustrates his later lectures are those in use today. Finally he explained quite correctly the apparent anomalies presented by the vapour densities of ammonium chloride, hydrogen sulphate, and mercuric chloride as due to dissociation in the vapour state.

These lectures must have been a beautifully clear piece of exposition aided by Cannizzaro's natural eloquence. However, the published account of them seems to have attracted little notice and he had to wait for the Karlsruhe Conference for his opportunity.

Early in 1860 Kekulé wrote to Weltzien, the professor of chemistry in Karlsruhe, about the possibility of calling an international chemical conference, and found him an enthusiastic supporter. Wurtz had also promised his support and so in March all three met in Paris to discuss the project with the French chemists, who agreed to support it. April and May were spent in correspondence with other colleagues and in June a letter, printed in English, French, and German, signed by forty-five of the leading chemists in each country, was widely circulated, inviting the recipients to attend an international conference at Karlsruhe on 3 to 5 September. The object of this conference was to try to get agreement on the following points: precise definitions of the ideas conveyed by the words, atom, molecule, equivalent, atomicity, basicity; examination of the true equivalents of substances and their formulae; initiation of a uniform notation and a rational nomenclature. Most of the replies were favourable and Weltzien went ahead with arrangements for the conference.

In the preliminary discussions about procedure with Weltzien and Wurtz, Kekulé opposed the appointment of a permanent chairman as the choice would arouse jealousy and the individual views of the chairman might carry too much weight. So it was agreed that a new chairman should be elected at each session. Kekulé mistrusted the older men and was anxious that the direction of the discussions should be in the hands of a strong secretariat of the younger generation. He was not in favour of inviting any prepared contributions, but hoped that general discussion would bring out the main issues that were in dispute and lead to some measure of agreement. He anticipated difficulties and was clearly afraid that the result might be inconclusive. Most of the prominent European chemists came to Karlsruhe and 140 were present at the opening meeting on 3 September.

Fortunately we have eye-witness accounts of the proceedings by Lothar Meyer,[2] Mendeleev,[3] and Savich[4] and also the long report written by Wurtz, one of the secretaries, including the text of Cannizzaro's speech on the last day.[5] All the documents at Karlsruhe relating to the conference were collected and indexed by Carl Engler when he wrote his account of it in 1892.[6] They were used later by Alfred Stock in a more detailed account including extracts from eleven of the interesting letters from Berthelot, Bunsen, Liebig, Pasteur, Roscoe, and Wöhler.[7] All these documents were destroyed by bombing during the last war but luckily Kekulé had kept a copy of Wurtz's précis which was reproduced in his biography by Anschutz.[5] This also contains Kekulé's notes for his opening speech and important details of the planning of the conference. So we know fairly accurately the course of events.

At the opening session Weltzien welcomed the members and took the chair with Wurtz, Roscoe, Kekulé, Strecker, and Schischkoff as secretaries. The notes of Kekulé's opening speech show that he failed entirely to give a clear directive to the conference. He rambled on, suggesting a variety of subjects that they might discuss going far beyond the limited objectives set out in the letter of invitation. He proposed that they should appoint a committee to formulate questions for discussion by the conference on the following day. This was agreed and a committee was appointed consisting of the secretaries and a number of members including Cannizzaro and Mendeleev with Kopp as chairman. It was agreed that the committee meetings should be private but Lothar Meyer says that additional members were co-opted until the committee included nearly half the conference.

The first meeting of the committee revealed a fundamental difference of approach by Kekulé and Cannizzaro as Kekulé would accept only chemical evidence for the determination of atomic and molecular weights and would not accept Cannizzaro's view of the identity of the physical and chemical molecule, pointing out the anomalies of the vapour densities of sulphur, ammonium chloride, and sulphuric acid. Eventually they decided on three questions for discussion:

1. Is it convenient to distinguish between atoms and molecules?
2. Can the term compound atom be replaced by the expressions radical or residue?
3. The idea of equivalents is empirical and independent of the idea of atoms and molecules.

At the second session of the Conference, Boussingault, one of the older generation, was in the chair, more familiar, as he said, with applied chemistry than with theory. He protested against the view that the conference was to seek to reconcile the old chemistry and the new, since, he said, chemistry does not grow old, but chemists do. Kekulé spoke to the first question and insisted that the magnitude of the chemical molecule could only be determined by chemical evidence and that physical methods were not acceptable, quoting sulphur as an example. Cannizzaro opposed this view saying that the distinction between the chemical and physical molecule was unnecessary and unsupported by evidence. The meeting adjourned without reaching any decision on this point. At the second meeting of the committee Kekulé and Cannizzaro had an argument about notation and Kopp, the chairman, said their business was to pose questions and not to discuss them in detail. He then spoke of the need to adopt symbols representing an agreed value. As it had been found necessary to double the atomic weights of certain elements he suggested the use of a barred atomic symbol to represent the doubled atomic weights and proposed the following question for discussion: 'Is it desirable to double certain atomic weights in order that the notation is in accord with recent progress?'

At the third meeting of the committee in the evening of 4 September Dumas took the chair as Kopp had to leave early. Kekulé proposed an alteration to Kopp's question and the chairman then interposed and spoke of the ill effects due to the present confusion, and the need to remedy it. He reminded them that twenty years ago the atomic weights of Berzelius were accepted by everyone and that nothing had happened to replace their authority. Wurtz supported this view taking advantage of the opportunity to point out Gerhardt's mistake in assigning to all metallic monoxides the same constitution as water. After a discussion it was agreed to put the following questions to the final session of the conference:

1. Is it desirable to make chemical notation conform to the progress of the science?
2. Is it acceptable to adopt anew the principles of Berzelius with certain modifications?
3. Is it desirable to distinguish by certain signs the new chemical symbols from those in use fifteen years ago?

No doubt Dumas's authority as the doyen of chemistry carried the day, but could the conference have met for its final discussion with a less auspicious programme? Dumas by his substitution theory had done more to discredit Berzelius than anybody, and now with the aid of Wurtz he is reinstating Berzelius with the object of discrediting Gerhardt. Savich, who like Cannizzaro was a supporter of Gerhardt, stigmatizes Dumas in his spirited account of the conference for his attitude to Gerhardt and Laurent at this stage, saying that he took advantage of his position as chairman to continue his attack on the dead men, which had had such tragic consequences in their lives.

At the final session Dumas was again in the chair and Odling had replaced Roscoe as one of the secretaries. According to Lothar Meyer and Mendeleev, Dumas opened the proceedings with a long speech, which Wurtz does not record. Developing the theme that there are two chemical sciences, the inorganic and the organic, he proposed that the older equivalent weights would be used in the former and the new in the latter. This roused Cannizzaro to the most vigorous opposition, deploring the habit of using different atomic weights in the two branches of the science and emphasizing its essential unity. In this he was later supported by Odling. He opposed the second question which sought to resuscitate the atomic weights and formulae of Berzelius which he said was illogical as much had happened to change them. He spoke at length of the value of Gerhardt's system, based on the work of Avogadro, Ampère, Dumas, and his pupil Gaudin. He pointed out the one inconsistency of Gerhardt's theory when he departed from the logic of Avogadro. Cannizzaro traversed most of the ground covered in his course of lectures and his hearers may well have had difficulty in following all his details though there is evidence that his speech drew general applause. His final words show that he did not expect to win complete support for Gerhardt's system, although, as he said, it was gaining new adherents every day among the younger chemists. He therefore advocated the use of barred symbols to express the double atomic weights in order to avoid the present confusion, although he must have seen that this was a retrograde step.

Strecker then pointed out the name of Berzelius had been substituted for that of Gerhardt in question 2 by a majority vote of the Committee, with which he disagreed. Kekulé, according to Wurtz, on this occasion supported Cannizzaro. Will reminded them that the object of the meeting was to agree on a clear logical notation that would not confuse young students of science. Erdmann then proposed that they should drop the first two questions as it was difficult to arrive at agreement on questions of principle or to impose a notation by a vote. Here Wurtz interposed to say there was no idea of voting on fundamental questions but simply on matters of formal notation. Odling with his usual clarity of mind insisted that at any rate they might agree that an element can have only one atomic weight and opposed the use of barred atoms which suggested that they were divisible. There followed a long discussion on the advisability of taking a vote on question 3. Cannizzaro opposed it, Kekulé was in favour, Kopp and others said that you cannot solve scientific questions by voting and that each investigator must retain his full freedom. Ultimately it was left to the chairman. Dumas then expressed the view that barred atoms representing atomic weights double those in previous use should be introduced. He then thanked the organizers and hosts of the conference and the meeting ended, as Lothar Meyer said, without reaching any conclusions, useful as it had been as an opportunity for the exchange of views. Stock comments that they must have parted in an icy mood.

No official record of the conference was issued although Wurtz's report had obviously been written for that purpose. Weltzien sent a copy to Kekulé who havered over it. He wrote in November saying that he had been too busy to attend to it. It was only a first draft and he wanted to adjust the differences it revealed between Cannizzaro's and his own 'unfortunate speech', but he did not see how it was possible to make similar adjustments in Dumas's speech.[5] He had obviously lost interest and was absorbed in his experiments on organic acids, so nothing was done.

However, that was not the end. The decisive moment had come after the close of the Conference when Pavesi distributed some reprints of Cannizzaro's *Sunto*. Lothar Meyer was given a copy which he put in his pocket to read on his way home.[2]

He wrote:

I read it again and again and I was amazed at the clear light which that little paper shed on the main subjects of our debates. The scales fell from my eyes, doubts disappeared and a feeling of certainty took their place. If I was able to help in clearing up the points at issue and cooling the hot tempers, I owe much to Cannizzaro's pamphlet. Many other members of the Conference felt the same. The tides of battle began to ebb; the old atomic weights of Berzelius once more came into their own. After the apparent discrepancies between the laws of Avogadro and Dulong and Petit had been explained by Cannizzaro, both could be used to the full and thereby the doctrine underlying the chemical values of the elements was put on a sound foundation without which the theory of atomic linkage could not have progressed.

It is thus certain that it was only after reading Cannizzaro's *Sunto* that Lothar Meyer realized the logic and clarity of his exposition. Without the convincing tables of values in the paper it must have been difficult to do justice to it verbally. How different the result might have been if Cannizzaro had had lantern slides or a hand-out. Mendeleev's enthusiastic praise of Cannizzaro in his letter of 7 September to his teacher Voskresensky must also have been written after he had read the paper as he quotes the tables in it.[3]

Two years later Lothar Meyer wrote his *Modern Theorien der Chemie*, published in 1864, which did much to clarify chemists' thinking about atoms and molecules, so that by 1870 Cannizzaro's views were generally accepted, except in France. Academician Figurovsky's biography of Mendeleev contains the following statement about the genesis of the Periodic Table, taken from some unpublished reminiscences by his son, D. I. Mendeleev:[8]

The decisive moment in the development of my theory of the periodic law was in 1860, at the conference of chemists in Karlsruhe, in which I took part, and at which I heard the ideas of the Italian chemist S. Cannizzaro. I regard him as my immediate predecessor, because it was the atomic weights which he found, which gave me the necessary reference material for my work. I noted immediately that the modifications he proposed to the atomic weights introduced a new pattern into Dumas' groupings, and it was then that I was struck with the essential idea of a possible periodicity in the properties of the elements on increase in the atomic weight. I was still hindered by the incongruities in the atomic weights accepted at this time; but I was firmly convinced that this was the direction in which to pursue my work.

So in spite of its inauspicious ending, the Conference, thanks to Cannizzaro's presence, was destined to have a decisive influence on the progress of chemical theory and to be a great landmark in its history.

I am grateful to Academician Figurovsky for his help and advice.

---

[1] CANNIZZARO, *Sunto di un corso di filosofia chimica* fatto nella R. Universita di Genova dal Prof. S. Cannizzaro. Tipografia Pieraccini, Pisa (1858).
[2] OSTWALD, Klassiker Nr 90. *Abriss eines Lehrganges der Theoretischen Chemie vorgetragen an der K. Universität Genua von Prof. S. Cannizzaro*

Herausgegeben von Lothar Meyer. Verlag Engelmann, Leipzig (1891).
[3] MENDELEEV. Letter to Prof. Voskresensky, in *Dimitrii Ivanovich Mendeleev, his Life and Works* by M. I. Mladentsev and V. E. Tischenko [Vol. 1, pp. 250–8]. Academy of Sciences, U.S.S.R. (1938).
[4] SAVICH, anonymous account of the Karlsruhe Conference in *Records of the Fatherland*, **134**, 77–82 (1861).
[5] ANSCHUTZ, R. *August Kekulé*, Vol. 1, pp. 183–209 and 671–691. Verlag Chemie, G.M.B.H., Berlin (1929).
[6] ENGLER, CARL, *Festgabe zum Jubiläum der vierzigjährgen Regierung Seiner Könglichen Hoheit der Grossherzogs Friederich von Baden*, pp. 346–355. Karlsruhe (1892).
[7] STOCK, ALFRED, *Der internationale Chemiker-Kongress Karlsruhe 3-5 September 1860 vor und hinter den Kulissen*. Zusammengestellt von Alfred Stock. Verlag Chemie, G.M.B.H., Berlin (1933).
[8] FIGUROVSKY, N. A. *Dimitrii Ivanovich Mendeleev*, pp. 44–51. Izd. Akad. Nauk. SSSR, Moscow (1961).

Among the many articles written about the Karlsruhe Conference the following are of special interest:

VON MEYER, E., Die Karlsruhe Chemiker-Versammlung im Jahre 1860, *J. prakt. Chem.*, n.s. **83**, 182 (1911).

DE MILT, CLARA, Carl Weltzien and the Congress at Karlsruhe. *Chymia*, **1**, 153 (1948).

DE MILT, CLARA, The Congress at Karlsruhe. *J. chem. Educ.*, **28**, 42, (1951).

9

HENRY ARMSTRONG
(1848–1937)
AND SOME OF THE GREAT FIGURES
OF NINETEENTH CENTURY
ORGANIC CHEMISTRY†

For more than half a century Henry Edward Armstrong was one of the dominating personalities among chemists. His devotion to chemistry, his belief in the power of scientific method, his wide outlook, and his gift of arousing the interest of young people made him a great teacher. He held strong opinions and never hesitated to express them. One soon learnt that there was an Armstrong 'doxy' on most things, which, exaggerated and extreme as it might sometimes seem, never failed to stimulate discussion. He kept his vitality to the end, and the flash of one of his brilliant waistcoats, like the flight of a kingfisher amid the sombre black of an evening gathering, was always a welcome sight to his friends. For it meant a warm greeting from his clear blue eyes, and a stimulating talk about whatever was uppermost in his mind at the moment. Armstrong always made one think, although he was fond of saying that very few of us were capable of doing so.

In writing of him I have in mind the opening lines of the Poet Laureate's poem on biography, so obviously his own; they should be a standing warning to all memorial lecturers.

> When I am buried, all my thoughts and acts
> Will be reduced to lists of dates and facts,
> And long before this wandering flesh is rotten
> The dates which made me will be all forgotten;

† The First Armstrong Memorial Lecture delivered before the Society of Chemical Industry at the Royal Institution on 21 November 1945.

And none will know the gleam there used to be
About the feast days freshly kept by me,
But men will call the golden hour of bliss
'About this time', or 'shortly after this.'

Men do not heed the rungs by which men climb,
Those glittering steps, those milestones upon time.
Those tombstones of dead selves, those hours of birth,
Those moments of the soul in years of earth.
They mark the height achieved, the main result,
The power of freedom in the perished cult,
The power of boredom in the dead man's deeds
Not the bright moments of the sprinkled seeds.

How Armstrong would have hated an orthodox memorial lecture about himself. And his memory is so fresh in the minds of his friends that it is difficult not to think of him as present with us. How much we miss his active, fertile brain. In his Frankland Memorial Oration, the lessons of Frankland's career set him thinking, and in the last half of the lecture all Armstrong's pet hobbies appear—the failure of science in education, coal conservation, the problem of nutrition, the value of pure milk, the need for a survey of the world's natural resources—particularly phosphates—ending with a doubt as to the quality of the essay that Members of the House of Commons would write on 'No Life without Phosphorus'. Unfortunately he singled out Mr. Churchill, whose essay we have good reason to know would not be lacking either in appreciation of science or clarity of expression.

There is no need to chronicle the events of Armstrong's life, as that has been done already in the understanding study of his life by his collaborator and friend, his eldest son.

I cannot speak with the advantage of having been one of his pupils, although I was in a sense a step-pupil. I was fortunate to learn chemistry at Dulwich from H. B. Baker, to whom I can never be sufficiently grateful for having taught me both by precept and example that life without research loses much of its zest and excitement. Baker was one of the few schoolmasters who, like Frankland, managed to do research with very slender opportunities. We soon realized that Armstrong was one of his heroes, whose encouragement meant much to him. Baker often went to Lewisham on Sundays to talk over his problems with Armstrong, and on Monday mornings we heard all about it. This made us feel that we were in touch with what was going on in the great scientific world outside.

When the time came for me to be fattened, as Armstrong put it, for the University Shows, Baker told me to buy Armstrong's article on inorganic chemistry in the ninth edition of the *Encyclopaedia Britannica*—and I have always kept it as a talisman, as Kolbe kept his first letter from Berzelius. Written in 1876, when Mendeleev's generalization was coming into prominence, it was almost the first consistent attempt to arrange the subject matter of chemistry in natural periods. I found it a most welcome relief from the formal textbooks, with its crisp, suggestive summary of each family of the elements, each illustrated with their more important heats of reaction.

I owe my first meeting with Armstrong, as I owe much else at Oxford, to Henry Miers's hospitality. It was rather a nervous moment for an undergraduate to meet Armstrong in his more leonine days. We knew he could be disconcerting. There was the story of the meeting at which there was a slight difference of opinion between the chairman and Armstrong as to the length and relevance of his remarks. When the next speaker began 'Mr. Chairman and Gentlemen,' Armstrong was heard to murmur 'A very proper distinction on this occasion.' But Armstrong forgot that I was one of the 'spoon-fed, over-taught, over-provided, over-examined, over-read and under-practised products' when he heard I was a pupil of Miers and Baker, had been in Munich with Groth and had made geological excursions with Zittel and Rothpletz. I had many a stimulating talk with him, and a happy friendship of more than forty years was unmarred by any clash, although sometimes Armstrong must have been suspicious of my fancy for ions. Once when I spoke to the Science Masters' Association on the Theory of Ionic Dissociation, with Armstrong, an ex-President, sitting in the front row, everyone was expecting a thunderstorm at my expense; I think I escaped because Armstrong knew I was not dabbling in 'fair hydrone' but in alcohol even drier than those vintages of which he was such a good judge.

We had many interests in common—crystallography, geology, scenery, and John Ruskin. I think Armstrong forgave

Ruskin for his hostility to modern science because of his love of crystals, his long study of the geological structure of scenery, and his beautiful drawings. I remember what pleasure Armstrong got from Pope's gift of Cook and Wedderburn's Library Edition of Ruskin, with its reproductions of so many of Ruskin's own drawings. He was a great admirer of Ruskin's style and of his fearless advocacy of the causes for which he cared. He thoroughly enjoyed the exuberance of Ruskin's purple passages, the eloquence of his defence of his heroes like Turner, and the invective with which he flayed the pretentious, the complacent, or the insincere. Armstrong, like Ruskin, would take infinite pains to find the right, the inevitable, word, and Ruskin had a genius for coining a word to emphasize his point. Of Sir John Lubbock's list of Best Hundred Books he wrote: 'Putting my pen lightly through the needless . . . and blottesquely,' as he did, 'through the rubbish and poison of Sir John's list—I leave enough for a life's liberal reading.' Armstrong probably felt a little envious, for he too was a great fighter.

The longest and most intimate talk I had with him was on a voyage to Morocco in 1936. It began by my asking him about the great chemists he had known in his younger days, and ended in a discussion of the history of organic chemistry when it was taking its modern form, which lasted until we reached Tangier. For me the talk was most revealing, as most of my impressions had come from Odling, an enthusiast for the French school. Gerhardt and Laurent had always been my heroes. Armstrong, a pupil of Frankland and Kolbe, belonged to the other camp. Kolbe had always been a puzzle to me. I didn't know the story of his partnership with Frankland, and that talk with Armstrong revived an old interest and gave me a much clearer picture of the contributions of the two schools.

For Armstrong the history of chemistry was not just a succession of theories, as it is so often painted. He was interested in the personalities of the chemists who made them, their characters, and their practical contributions to the corpus of chemical knowledge. In the period we discussed, the efforts of almost all the outstanding chemists were directed to organic chemistry. It was a period of rapid growth, when there were great discoveries to be made by individuals. Each had his own theory of which he was tenacious, as it meant much to him; it was in a sense the

Henry Armstrong

scaffolding which enabled him to make his individual contribution to the structure of chemistry. And Armstrong judged not by the theory but by the results. 'By their fruits ye shall know them.' And was it not Armstrong who said that 'Hypotheses like Professors, when they are seen not to work any longer in the laboratory, should disappear'?

The best landmarks of progress in these years are the great textbooks, not those written by compilers such as Gmelin, useful as they were, but those written by the men who were reshaping organic chemistry in the laboratory, and felt the urge and had the energy and imagination to give their picture of the whole range of the subject. These milestones were the third edition of the great *Lehrbuch* of Berzelius translated by Wöhler in the years 1833 to 1841, in which organic substances are classified under their various plant or vegetable origins; Liebig's textbook of 1839 which opens with a brave phrase, echoing Lavoisier, 'Organic chemistry is the chemistry of compound radicals', but after starting with compound radicals as a basis of classification he soon relapses into the Berzelian system; Gerhardt's *Précis* of 1844, using his own classification based on homologous series, his *Introduction to Chemistry by the Unitary Method* of 1848, and the four volumes of his *Traité* of 1853–6; Kolbe's textbook of 1854–60, written with his shrewd insight into the structure of organic acids; and finally Kekulé's unfinished *Lehrbuch* of 1859–67 in which organic chemistry began to take its modern form. Here they are as a visual record of Armstrong's conversation during those three unforgettable days.

He began with Laurent and Gerhardt, the two young rebels who saw the weakness of the Berzelian system, and gave their lives to remodelling organic chemistry and giving it a scientific classification based once again on the clear logic of a French mind. Here I was able to contribute something from my many talks with Odling. Armstrong admired the courage and tenacity with which Gerhardt and Laurent fought tradition and authority, fearless of the effects on their personal fortunes, but he deplored the completeness of their ultimate victory which he felt had obscured progress by preventing a full appreciation of the work of Frankland and Kolbe. For me theirs is one of the epic partnerships, to be ranked with Lavoisier and Laplace, Liebig and Wöhler, and I would now add Kolbe and Frankland.

They suffered almost persecution for their views, and Laurent died as a result of great privation. Their laboratory researches were carried out under incredible difficulties and lack of support but they had the unquenchable fire of devotion to chemistry.

Laurent, the elder by eight years, was the son of a peasant. Trained as a mining engineer, in 1831, at the age of twenty-four, he was Dumas's lecture assistant, and was working with him on naphthalene and its derivatives. Armstrong was interested in him as one of the earliest chemical crystallographers, for besides preparing a large number of new substitution derivatives of naphthalene he measured their crystalline form and showed that they formed an isomorphous series. This gave added support to the view of the French chemists that in these substitution products chlorine occupied the same position as the hydrogen it replaced. It was probably also the basis of Laurent's theory of nuclei, in which he pictured a parent structure with carbon atoms at the angles of a rectular polyhedron and hydrogen atoms at the middle of each side replaceable by other groups, with the possibility of additive compounds corresponding to pyramids on the faces. Laurent's nuclei were comparable with Dumas's types, with the added hypothesis of the geometrical structure of the molecules.

Laurent is remembered for his association with Gerhardt and the important influence he had on Gerhardt's views, but he was a most skilful organic chemist, and he added greatly to our knowledge of many groups of organic compounds.

Gerhardt, born in 1816, was the son of an Alsatian chemical manufacturer. He published his first paper, on the classification of silicates, when he was eighteen, and finding the routine of his father's business intolerable, with the help of friends he spent two years with Liebig at Giessen. Returning home he quarrelled once again with his father and set out, at the age of twenty-three, to seek his chemical fortune in Paris, armed with a letter to Dumas from Liebig and 200 francs in his pocket. He kept himself by doing literary work, including the translation of Liebig's textbook, which gave him a wide view of the whole range of organic chemistry as it then was.

Right from the outset Gerhardt seemed to realize the need for some entirely new approach to the problems that were presented by the rapidly growing number of organic compounds. In most of his papers there is an attempt at a generalization, often based, so his opponents said, on insufficient evidence, but Gerhardt had the same sort of vision of a new system as Lavoisier, and each of his main papers marked a step towards the reorganization of organic chemistry at which he aimed.

By 1840 the battle over substitution had been won and the authority of Berzelius had begun to be seriously challenged. For Berzelius as for Lavoisier oxygen was the central element. His massive contributions to chemistry had been built up on Lavoisier's dualistic conception of chemistry, reinforced by the electrochemical theory of chemical affinity. The elements of organic chemistry, for Berzelius as for Lavoisier, were the compound radicals, into whose composition oxygen could not enter, as by definition a radical is an oxide less oxygen. Thus Berzelius wrote the formula of acetic acid $C_4H_6O_3.H_2O$, a combination of the oxide of the radical C_4H_6 with water, which in the acetates was replaced by metallic oxides. That electronegative and electropositive elements like chlorine and hydrogen could occupy the same place in chemical compounds was unthinkable for Berzelius.

But now dualism, the electrochemical theory, and Berzelius's conception of radicals had received a severe blow. Even Berzelius's atomic weights were being replaced by those of Gmelin in which the values for carbon and oxygen were halved to make them 6 and 8 respectively. Dumas, Laurent, and others had shown that electronegative elements like chlorine could replace hydrogen atom for atom in organic compounds with little change in properties and, as Laurent showed, without change of crystalline form. Berzelius fought hard against the idea that such a replacement could take place, and assumed that the substitution was accompanied by a change of structure. He gave acetic acid the formula $C_4H_6O_3.H_2O$ and trichloracetic acid $C_2Cl_6.C_2O_3.H_2O$. But the facts were against him, and in 1840 he seized on Gerhardt's conception of bodies like nitrobenzene as conjugate compounds formed by the union of two residues with the elimination of water, the organic radical being called by Gerhardt a 'copula'. Berzelius's formulae now became conjugate, and substitition was admitted in the copula. Acetic acid was written as $C_2H_6.C_2O_3.H_2O$ and trichloracetic acid $C_2Cl_6.C_2O_3.H_2O$. And so conjugate compounds and copulae, Berzelius' practical admission of defeat, took their place in chemical literature for twenty years. In the hands of Kolbe they were made the basis of an important contribution to structural chemistry.

Gerhardt's paper on nitro-compounds which started the idea of conjugate compounds and copulae was followed by work on cymene derivatives which gave him the notion of homologous series of organic compounds. This was the basis of a revolutionary paper in 1842 on the classification of organic substances. Gerhardt now saw the need to ensure that formulae should represent comparable magnitudes. These had been based previously to a large extent on the theory of dualism which he rejected. Following Ampère he insisted that the weights of equal volumes of vapour must represent comparable magnitude and he first adopted four-volume formulae referred to one volume of hydrogen as unit. But later, realizing that inorganic formulae were on a two-volume basis, he made this the basis of his organic formulae, and he doubled the Gmelin equivalents for carbon and oxygen, thus arriving at practically the same values of molecular and atomic weights as we use today.

In 1844 he published the first volume of his *Précis de Chimie Organique*, using the homologous series as the basis of classification together with the chemical families of compounds which contain the same number of carbon atoms and can be derived from one another. Gerhardt having lost faith in dualism, the electrochemical theory, and the Berzelian radicals, turned to a purely empirical outlook, thus avoiding the hypothetical formulae of other chemists, which he regarded as figments of the imagination. He used empirical formulae based on a two-volume basis and he used the word 'equivalent' indiscriminately for atomic, molecular, and equivalent weights.

The new views roused a storm of opposition from the older chemists, and Berzelius did his best to discredit them in the *Jahresbericht* in which he had reviewed the current chemical literature since 1822. Gerhardt, undaunted by this formidable opposition, determined to get a hearing for his own point of view. In 1845 he founded a new journal, *Comptes Rendus mensuels des Travaux Chimiques*, which contained both original papers and a summary by himself of current literature which gave him the opportunity of replying to Berzelius. It was a courageous venture in view of Gerhardt's slender resources, but with his indomitable energy he produced the new journal, completed the second volume of the *Précis*, and carried on his laboratory researches single-handed.

In 1844 Gerhardt and Laurent had decided to collaborate. It was a fortunate and stimulating partnership, for Laurent like Gerhardt was seeking a logical basis of classification along rather different lines, and Gerhardt's impetuosity gained much from Laurent's better balanced judgement. Gerhardt had gone too far in his reaction against the older theories, in his use of purely empirical formulae and his neglect of any distinction between atoms and molecules. Laurent pointed this out to him, and Laurent's own constructive contribution to the new system came in 1846 in a paper on the organic compounds of nitrogen. In this he supports Gerhardt's two-volume formulae, and points out that in these formulae the sum of the atoms of hydrogen, nitrogen, and the halogens is always an even number. The examination of apparent exceptions to this rule showed that they were based on incorrect analyses, and numerous revisions of formulae were suggested.

Laurent then applied the two-volume formulae to elementary gases, and showed that their molecules must contain two atoms, which led him to the modern definitions of atom, molecule and equivalent, and to the explanation of nascent action, which until then had been a mystery.

In 1848 came Gerhardt's final blow at dualism in his *Introduction a l'étude de la Chimie par le système unitaire*, in which he classes substances both according to their chemical functions and in homologous series. He uses unitary formulae, and is still sceptical about Laurent's definition of atoms and molecules. He uses the term 'proportional number' for both.

Gerhardt's violent reaction against the welter of formulae was natural. For most of them had no practical justification, and in his eyes they merely darkened knowledge. But Laurent now convinced him that he had gone too far in using only empirical formulae, as there certainly were groups of atoms which were transferred from one compound to another in chemical reactions and Gerhardt had already admitted this in his conception of residues and copulae. It was along these lines that their so-called synoptic formulae were developed.

Their collaboration at this time was difficult, as Gerhardt was in Montpellier and Laurent in Bordeaux, and they met only once. They were eager to be working together in Paris, but the opposition to their views was intense and they were regarded almost as dangerous criminals by the older chemists who controlled all chemical appointments. When Laurent applied for posts, on two occasions they were given to Balard.

They criticized Liebig's work on the derivatives of melanine, and particularly his analysis and formula of mellone. Liebig was furious and published a tract 'M. Gerhardt et la Chimie Organique' in which he called Gerhardt a highway robber and accused him of stealing other people's belongings and using them himself. Liebig also tried to break up the partnership by telling Laurent that Gerhardt was 'un homme sans moralité' and others warned Laurent that his association with Gerhardt would ruin his prospects.

But Laurent's loyalty to Gerhardt was superb. He was now in Paris without an appointment, living by giving lectures and often in great want, but he never hesitated to take Gerhardt's quarrel on his own shoulders. There is a pathetic letter to Gerhardt saying that he has no alcohol or ether and must stop work unless Gerhardt can authorize him to buy 20 francs' worth of chemicals on his account. In another, he says, 'Poverty has just made me commit a crime—I have sold Masson a bad book on crystallography and the blowpipe for 500 francs.'

In 1848 Gerhardt got a year's leave from Montpellier without salary in order to be with Laurent in Paris. In the following year, as no appointment was forthcoming, and Laurent had lost a small post he had at the Mint owing to ill-health, Gerhardt borrowed some money and with Laurent opened a small laboratory where they could work together and make a living by taking pupils. Meanwhile their views were steadily gaining adherents, particularly in England, where Williamson was a stout friend and supporter.

Then came a series of discoveries in other laboratories which had a decisive effect on the development of Gerhardt's views. In 1849 Wurtz prepared the primary amines by the hydrolysis of isocyanates. In the following year Hofmann isolated the primary, secondary, and tertiary amines by the action of ammonia on alkyl iodides, and in the same year came Williamson's historic paper on etherification and his synthesis of ethers by the action of alkyl iodides on the alcoholates of potassium.

The formulae of alcohol and ether and their chemical relationship had been in hot debate for a quarter of a century, and Williamson now established the correctness of Gerhardt's view of their relative molecular magnitudes. He wrote their formulae as derivatives of water by the successive replacement of hydrogen by ethyl thus:

$$\left.\begin{array}{l}H\\H\end{array}\right\}O \qquad \left.\begin{array}{l}C_2H_5\\H\end{array}\right\}O \qquad \left.\begin{array}{l}C_2H_5\\C_2H_5\end{array}\right\}O$$
$$\text{Water} \qquad\qquad \text{Ethyl alcohol} \qquad \text{Ether}$$

Odling often told me that his contemporaries were never convinced about physical methods for determining molecular weights, but they accepted Williamson's chemical proof of the relative magnitudes of alcohol and ether as decisive. His notes for a lecture in 1899 on 'Chemical Theories under Discussion about the year 1850' contain the following passage: 'Dependency then of determination of atomic weights on possibility of determining unit weights—means for determination of unit weights. Conclusions from gas or vapour densities and other physical considerations. Deduction of molecular or unit weight scarcely accepted by physicists and not at all by chemists. Necessity for determination of unit-weights on chemical grounds for establishment of concordance of results with those based on physical grounds.'

Gerhardt was no mere theorist as his enemies liked to make out. During these years in Paris, in spite of their difficulties, he and Laurent had a big output of research and the decisive moment came for Gerhardt with his synthesis of the anhydrides of monobasic acids in 1852. Hitherto, apart from Deville's preparation of nitric anhydride, the only anhydrides which had been isolated were those of dibasic acids such as sulphuric and succinic in agreement with Gerhardt's formulae, although dualistic formulae made the existence of monobasic anhydrides equally probable. Gerhardt by the reaction of acid chlorides with the alkali salts of monobasic acids succeeded in preparing a whole series of anhydrides. It was a parallel to Williamson's work on the ethers, when the method of preparation was conclusive evidence of their formulae.

Not only was Gerhardt's work a final proof of the untenability of the dualistic formulae for monobasic acids, but the similarity with the ethers set him thinking, and the paper contains the first statement of his theory of types in which he referred the formulae of organic compounds to four types, water, hydrogen, hydrochloric acid, and ammonia. By the substitution of various organic groupings for hydrogen the different classes of organic compounds could all be represented as derived from these simple inorganic molecules, thus giving a new method of classifying them and showing the relationships of their chemical composition. Gerhardt's type formulae were an unconscious recognition of the different combining powers of the atoms and groups, and led inevitably, I had always thought, to Kekulé's theory of atom linkage.

At this point, when we were discussing the influence of Williamson on Gerhardt, I fortunately asked Armstrong a question that he evidently regarded as very ill-informed. Forgetting for a moment the baptismal names of his first-born and his own early masters, I asked whether he thought that Williamson or Frankland had made the greater contribution to chemistry. I was quickly left in no doubt as to his opinion, and only hunger ended his recital of the Kolbe–Frankland saga. No doubt the story was told in full in his Frankland Memorial Lecture, but the MS. of this was unfortunately lost so I must try and reconstruct it from what Armstrong told me.

Till then Kolbe had always been an enigma to me, as I daresay he is to others. It seems so difficult to reconcile his queer formulae, his apparent conservatism, and his violent attacks on Kekulé and van 't Hoff, with his success as a teacher and his great contribution to structural chemistry. Born in 1818, the son of a clergyman, Kolbe's training was under Wöhler at Göttingen, where he met Berzelius and was no doubt brought up by Wöhler in the true Berzelius tradition. His early work with Wöhler was on the chlorination of carbon disulphide to give thiophosgene and carbon tetrachloride. From the former he prepared a series of chlorinated methylsulphonic acids which, following Berzelius, he formulated as conjugate compounds with the radical methyl acting as a copula. Berzelius was delighted at the discovery of a series of analogues to the chloracetic acids. He devoted six pages of the *Jahresbericht* to Kolbe's paper, and won his heart by writing him an encouraging letter which Kolbe kept as a talisman all his life. There is little doubt that the success of this investigation coloured the whole of Kolbe's life's work and outlook.

For Gerhardt, rational formulae were only a convenient means of representing the reactions in which a substance might take part, and the same substance could have more than one rational formula. Even Kekulé as late as 1861 still kept this Gerhardtian view. But for Kolbe the conjugate formulae of Berzelius, adapted as we have seen from Gerhardt, meant something real and he set out to try to prove their truth both by analysis and synthesis. Encouraged by the early success of his experiments he never wavered from this view, and this explains both his conservatism and the motive power behind his researches. Kolbe had also obtained trichloracetic acid by passing carbon tetrachloride through a heated tube, and hydrolysing the resulting tetrachlorethylene in presence of chlorine. This was one of the earliest examples of the synthesis of an organic compound from its elements. Most of this work was done in Marburg, where Kolbe had gone in 1842 as Bunsen's assistant, and had learned his methods of gas analysis. In 1845 he came to London as assistant to Lyon Playfair, to carry out the analysis of mine gases for the Commission that was sitting on explosions in coal mines. There he met Frankland, who had just been appointed by Playfair as junior assistant to undertake the analysis of minerals at the School of Geology in Jermyn Street.

Frankland, who was then twenty years old, had begun his chemical career, like Dumas, as a druggist's apprentice. This had taught him little, but a Lancaster surgeon named Johnson, realizing the lack of training of apprentices like Frankland, had made a small laboratory and lecture room to encourage them, and took an active interest in their studies. Struck by Frankland's keenness and ability he got an introduction for him to Lyon Playfair, and then, as Armstrong said, Frankland began 'to learn chemistry on the Squeer's principle by making it'. Within two years, with the help of Kolbe, he was making chemical history.

Kolbe's heart was in organic chemistry, not gas analysis. He wanted to discover the groupings in an organic compound, to prove the existence of the radicals which Berzelius called copulae. His first success came when he isolated the butyl-radical (actually octane) by the electrolysis of a strong solution of valeric acid. He quickly interested the young Frankland in organic chemistry, and together they began a research to try to establish the molecular structure of the fatty acids by synthesis. They prepared ethyl cyanide and on hydrolysis it gave, as they hoped, propionic acid (they called it metacetonic acid), thus proving the existence of the ethyl radical in the acid. Once again the success of a first research was the turning point in a career. When Kolbe returned to Marburg he took Frankland with him to continue their joint research and to learn gas analysis from Bunsen. They first extended their work to methyl and amyl cyanides which gave, as they expected, acetic and caproic acids on hydrolysis. They then tried to isolate the radical ethyl by the action of potassium on ethyl cyanide; they failed to do this but discovered an interesting polymer of ethyl cyanide and investigated its derivatives. After three months in Marburg, Frankland had to return to England to teach chemistry at Queenwood College, a school in Hampshire, where Tyndall was his colleague. There he continued his attempts to isolate hydrocarbon radicals by heating ethyl iodide in sealed tubes with potassium, when he got a mixture of ethylene and ethyl hydride. The experiment was repeated using zinc in place of potassium, but a violent explosion during the analysis of the resulting gas shattered Frankland's eudiometer, and his experiments were held up until he went again to Marburg with Tyndall in the autumn of 1848. Analysis then showed that he had isolated what he called ethyl, which was in fact butane. He continued the work with ethyl and methyl iodides, studying the reactions of the halogens with the hydrocarbon radicals until, on 12 July 1849, in Marburg, came the dramatic discovery of zinc methyl, the first of the organo-metallic compounds. Frankland had added water to the solid residue left after heating methyl-iodide with zinc, when a greenish blue flame several feet long shot out of the tube causing great excitement in the laboratory. Bunsen was at first alarmed lest it should be cacodyl, due to an impurity of arsenic in the zinc, but it was quickly found that the metallic constituent was zinc, and the formula of the new substance proved to be $Zn(CH_3)_2$.

Shortly afterwards Frankland was invited by Liebig to spend a semester at Giessen, and there he isolated the radical amyl and began to work on zinc amyl. At the end of the year he returned to England to take Playfair's place as Professor at the Civil Engineering College at Putney until 1851, when he went as the first professor of chemistry to the newly founded Owens College in Manchester, where he remained until 1857.

Frankland was quick to see that his discovery of zinc methyl opened up a wide field of research, and he first prepared a number of the alkyl compounds with zinc and tin and later with mercury and boron and made a most careful and systematic study of their properties. He soon realized their importance as synthetic reagents—zinc methyl, he said, 'will be capable of replacing electronegative elements in organic or inorganic compounds by ethyl—a kind of replacement which has never yet been attempted, but which the author anticipates will enable him to build up organic compounds from inorganic ones and ascend the homologous series of organic bodies'. Frankland was the pioneer in using the organo-metallic compounds in organic syntheses for the next twenty years.

Armstrong often spoke of Frankland's extraordinary practical ability—'I have never met with a more skilful worker'—and those early papers of Frankland's in a new and most difficult field bear witness to his great experimental skill and originality. He was a great thinker too, and within three years of his discovery of zinc methyl his study of its analogues led him to the first general statement about the combining powers of the elements, afterwards known as the theory of valency.

Zinc methyl was similar in constitution to Bunsen's cacodyl which Kolbe regarded as a conjugate compound of arsenic with methyl as a copula. Now according to Berzelius the presence of a copula should not alter the combining power of an element or group: acetic acid for instance had the same basicity as formic acid in spite of the presence of the methyl copula. Cacodyl like arsenic combined with oxygen and the halogens and formed an acid. Hence Frankland expected zinc methyl to behave similarly and was surprised to find that it had no combining power, merely exchanging its hydrocarbon groups for others. A study of the properties of stannous and stannic ethides, and a comparison of the formulae of cacodyl and organic antimony compounds convinced him that elements had a definite combining power and that the attachment of each methyl group reduced by one their power to combine with other elements. '*No matter what the character of the uniting atoms may be, the combining power of the attracting element*, if I may be allowed to use the term, *is always satisfied by the same number of these atoms.*'

The paper containing this generalization which was to transform chemical theory was communicated to the Royal Society in May, 1852, but was not published for twelve months as it was mislaid by the Secretary (Professor, afterwards Sir Gabriel Stokes), and Frankland thought it had not been thought worthy of publication. A delay at this critical moment was most unfortunate as, to quote Armstrong, 'the problem first solved by Frankland was in the air—chemists everywhere had it in mind, especially in France'. Gerhardt's types were a partial recognition of the same principle, extended by Williamson to multiple types with polyvalent radicals, the valency or atomicity of which was first indicated by Odling by the familiar dashes. But Frankland was the first to see that there is a general law of atomicity or valency underlying chemical combination.

Meanwhile, Kolbe had left Marburg in 1851 to spend three years in editing Liebig's and Wöhler's *Dictionary of Chemistry* for Vieweg, and in 1854 the first part of his own *Textbook of Organic Chemistry* was published. In the Introduction he makes his own position quite clear. He still regards the radical theory as the safest basis for teaching chemistry and says that he belongs to the conservative party of chemists. He attacks Gerhardt and Laurent for their unsound generalizations, and for the type theory. 'Chemistry is something more than a mere arithmetical exercise into which Laurent and Gerhardt think they can convert it.' He attacks Williamson's and Gerhardt's formulae for alcohol and ether on the ground of the products of electrolysis of alcoholates. He refuses even to accept Frankland's ideas on atomicity, as he would not admit that electronegative elements could be replaced by electropositive radicals such as methyl and ethyl. His formulae are based on dualism using Berzelius's copulae and Gmelin's equivalents. Not, it would seem, a promising beginning! But Kolbe's strength was in the laboratory and he collects for the first time in one chapter all the methods of preparing derivatives of organic compounds by means of chemical reagents, which have since become standard: oxygen and oxidizing agents, reducing agents, halogens, phosphorus pentachloride and oxychloride, nitric and nitrous acids, and sulphuric acid. His classification of compounds is partly by radicals and partly by acids and their derivatives, to our knowledge of which he was to contribute so much.

The difference of opinion between Kolbe and Frankland was soon settled by correspondence between them. Kolbe accepted Frankland's theory and in 1857 they published a paper† 'On the Constitution of the Fatty and Aromatic Acids, Aldehydes, Ketones, etc., and their Relation to Carbonic Anhydride', giving their joint view that the oxygen atoms in oxides of metals and non-metals can be replaced by the same number of atoms of a positive atom or radical, the replacement being accompanied by an increase in the basic properties of the compound. Thus Frankland had an important influence in changing Kolbe's outlook and even before their joint paper Kolbe was teaching that organic compounds can be regarded as derived from carbonic acid by the replacement of oxygen atoms by radicals. In this he was following a suggestion made by Liebig in 1847 in a paper on the chemical processes underlying respiration. Kolbe elaborated the theory in 1860 in his classical paper on 'The Natural Relation between Organic and Inorganic Compounds', and it was the basis of his pioneer work which revealed the structure of so many organic compounds.

Using atomic weights of 6, 8, 16, and 35·5 for carbon, oxygen, sulphur, and chlorine respectively he first wrote formulae as follows:

Dibasic carbonic acid	$2HO \cdot C_2O_4$
Monobasic methyl carbonic or acetic acid	$HO \cdot C_2(C_2H_3)O_3$
Acetic aldehyde	$C_2(C_2H_3)HO_2$
Acetone	$C_2(C_2H_3)_2O_2$

Then realizing that the basicity of the acid was related to only

† Frankland's name was accidentally omitted as a joint author with Kolbe, but the paper throughout uses the pronoun 'we'.

two of oxygen atoms in C_2O_4 since it vanished when these two were replaced by other atoms or radicals, he distinguished between the oxygen inside and outside the radical $[C_2O_2]$ to which he gave the name carbonyl:

Carbonic acid	2HO.	$[C_2O_2], O_2$
Formic acid	HO. H	$[C_2O_2], O$
Acetic acid	HO. (C_2H_3)	$[C_2O_2], O$
Acetaldehyde	$\left.\begin{array}{c} C_2H_3 \\ H \end{array}\right\}$	$[C_2O_2]$
Ethyl alcohol	HO.$\left\{\begin{array}{c} C_2H_3 \\ H_2 \end{array}\right\}$	C_2, O
Unknown alcohols	HO.$\left\{\begin{array}{c} C_2H_3 \\ C_2H_3 \\ H \end{array}\right\}$	C_2, O
	HO.$\left\{\begin{array}{c} C_2H_3 \\ C_2H_3 \\ C_2H_3 \end{array}\right\}$	C_2, O
Propionic acid	HO.$[C_4H_5]$	$[C_2O_2], O$
Lacetic acid	HO.$\left(C_4\left\{\begin{array}{c} H_4 \\ HO^2 \end{array}\right\}\right)$	$[C_2O_2], O$
Glyceric acid	HO.$\left(C_4\left\{\begin{array}{c} H^3 \\ HO_2 \\ HO_2 \end{array}\right\}\right)$	$[C_2O_2], O$

These formulae, clumsy as they may seem to us, were based on Kolbe's own study of organic compounds and their synthesis. They had a very real significance for him and by comparison he considered Gerhardt's type formulae superficial and lacking in any fundamental basis. Having always two atoms of carbon in place of one must have made it difficult for him to appreciate the theory of atomic linkage, but he used his formulae with an uncanny instinct to predict the existence of unknown compounds, such as the secondary alcohols discovered by Friedel in 1862 and the tertiary alcohols by Butlerow in 1864, using Frankland's zinc methyl as a synthetic reagent. But more important still was the use he made of them in directing the research in his own laboratory to the elucidation of the structure of more complex organic substances. Outstanding examples of this are: the reduction of malic and tartaric acids to succinic by Schmitt in 1860, Kolbe's own researches on the constitution of lactic acid and alanine, and his synthesis of taurine in 1862, Volhard's synthesis of sarcosine in the same year, von Oefele's discovery of the sulfine and sulfone compounds in 1864, Kolbe's synthesis of malonic acid from cyanacetic acid in the same year, and finally his synthesis of salicylic acid and his discovery of nitromethane almost simulataneously with Victor Meyer.

While Kolbe was making these outstanding contributions to the synthesis of organic acids and to our knowledge of their constitution, Frankland was also active in the same field. In 1857 he left Manchester to get more time for research at St. Bartholomew's Hospital and in 1863 he succeeded Faraday at the Royal Institution.

With the help of zinc alkyls Frankland, working with Duppa, synthesized a number of new members of the lactic acid series from which they prepared the corresponding acrylic acids and their derivatives. Finding that when ethyl acetate is heated with sodium a large volume of hydrogen is evolved they determined to utilize this reaction for the synthesis of higher members of the fatty acid series. In the course of this work they isolated acetoacetic ester independently of Geuther and used it to synthesize a large number of fatty acids, ketones, and ketonic esters.

Thus both Kolbe and Frankland, working outwards from their early joint researches on independent lines, made a great contribution to synthetic chemistry just at the time when the theory of atomic linkage enabled such knowledge to be applied to the structure of organic chemistry. When Duppa died in 1873 Frankland's interest in organic chemistry seemed to lapse, but Kolbe was active until his death in 1884.

When the Kolbe–Frankland saga was ended Armstrong and I went back to where we had left Gerhardt and Williamson and the type theory in 1853, when Kekulé, the maker of modern organic chemistry, was just coming on the scene.

Gerhardt's types had an immediate success with the younger chemists. Williamson at once extended them to include multiple types to cover the case of polyatomic radicals and polybasic acids:

$$\left.\begin{array}{l}SO_2\\H_2\end{array}\right\}O_2 \qquad \left.\begin{array}{l}PO'''\\H_3\end{array}\right\}O_3$$

Odling introduced mixed types and indicated the atomicity of the radicals by the well known dashes.

It was tragic that Laurent should have died in 1853 just when the long fight was practically over and Gerhardt's system to which he had contributed so much was winning general acceptance. Laurent had sacrificed everything for science and had achieved so much in the face of great difficulties. He died exhausted by the long struggle against poverty and against almost fanatical opposition.

In 1854 Gerhardt tasted success. He was appointed to a professorship at Strasbourg, where he was to build a new laboratory, and by the irony of fate he was asked to write the new edition of Berzelius' *Organic Chemistry*. He accepted on condition that he could use his own classification in homologous series and chemical families, but at long last he had learnt the value of compromise. Only in the final chapters on Generalities does he use his type formulae and his own atomic weights; in the body of the work he uses Gmelin numbers and even gives the dualistic formulae. When Pebal asked him why he had not used his own clearer presentation throughout, he laughed and answered, 'Then nobody would have bought my book!'

In the final chapters he expounds his type theory, and shows how the formula of organic substances can be derived from four types by the substitution of hydrogen by radicals or residues. Radicals he defined as groups of atoms which can be exchanged in double decompositions but cannot be isolated. 'Chemical formulae are not intended to represent the arrangement of the atoms in a molecule, but their object is to show in the simplest and most exact way the relation between substances.' Gerhardt regarded his formulae as contracted chemical equations, and while therefore substances might have more than one formula, most could in fact be represented by only one.

These chapters were his last legacy to science, as he died suddenly in 1865, when he was correcting the proofs, at the age of 40.

Gerhardt's interest was in classification, in finding a system within which the rapidly growing multiplicity of organic compounds would find their logical place, thus making possible the scientific study of their properties and relationships. He was sceptical as to the possibility of ever knowing the actual atomic structures of molecules. His formulae were formulae of convenience, but the progress he had made and the order he and Laurent had introduced were destined very quickly in the hands of Kekulé to give organic chemistry its modern form.

Kekulé after studying chemistry under Liebig from 1849 to 1851 spent a year in Paris, where he met Gerhardt who gave him the manuscript of his treatise to read. This left a lasting impression on him, and as Armstrong said 'he was enthralled by Gerhardt'. In 1853 he came to London as assistant to Stenhouse who was then Chemist to the Mint. In London he saw a great deal of Williamson and Odling, both enthusiasts of the Gerhardt school, and he always said that his structural theory was born in his dream about atoms on the top of an omnibus between Islington and Clapham Road, where he lodged. In 1856 he went to Heidelberg as *privatdozent*, and his investigations of the constitution of mercury fulminate led him in 1857 to recognize the tetravalency of carbon and to add to Gerhardt's types the marsh gas type. In 1858 he developed the idea at length in a paper on 'The Constitution and Metamorphosis of Chemical Compounds and the Chemical Nature of Carbon', and the first parts of his *Lehrbuch* were published in 1859.

Kekulé recognizes at the outset that the hydrogen, water, and ammonia types, to which he had added the marsh gas type, are only the recognition or expression of the combining powers of the elements or radicals, and he uses Odling's dashes to indicate their atomicity or basicity or what we call valency—but he made no reference to Frankland. His rational formulae are based on Gerhardt's types, and he explains in almost the same words as Gerhardt that these rational formulae are only reaction formulae, not constitutional formulae, and that they do not express the relative positions of the atoms in the compound. He emphasizes this 'as some chemists seem to think that they can determine the constitution of compounds from their reactions'. In any case, says Kekulé, this would need a perspective drawing and cannot be shown by the arrangement of atoms in one plane. Chemistry, he said, will never reveal the structure of the molecule, but possibly physics may.

When Kekulé comes to consider the carbon compounds in detail his great advance was the recognition of the formation of carbon chains with either single or multiple links between the carbon atoms, but he did not seem to realize the power of the instrument he had constructed. He wrote a few graphic formula for two carbon bodies, such as acetic acid, which show accurately their constitution for the first time on a valency basis. He realizes that the properties of the atoms are dependent on their relative position in the molecule, and yet he does not apply these formulae to substances containing three carbon atoms when they would have explained the isomerism of the lactic acids, as he says that they cannot give a true picture of the positions of the atoms. The isomerism of ethylene dichloride and ethylidene dichloride is left unsolved when the solution is ready at hand if he had used his graphic formulae consistently. He criticizes the use of the formulae used by Kolbe in which the difference between alcoholic and acidic hydrogen is explained. 'Such formulae have no advantage', he says, 'over the type formulae, they conceal a number of analogies and in other cases suggest analogies where none exist.'

All this shows that Kekulé was suffering from what Armstrong called 'Gerhardtism', and that at this time he did not realize the full value of the new instrument he had put into chemists' hands. Gradually its value dawned on him, and he used it with greater confidence. The first volume of the *Lehrbuch* was completed in 1861, the second volume on aromatic compounds based on Kekulé's benzene ring appeared in 1866, and the third volume was never finished. Kekulé gave organic chemistry its modern form and fifty years later physical analysis, as he had predicted, established the reality of the formulae as to which he was at first so sceptical.

Kolbe never forgave Kekulé for his advocacy of Gerhardt's type theory, for his failure in the *Lehrbuch* to make any recognition of Frankland's contribution to the theory of valency or atomicity, and for his rather scornful reference to Kolbe's own work. For twenty years he lost no opportunity of attacking Kekulé's views and giving Frankland the credit that was his due. It is a curious example of how far personal ties can warp the judgement of one who had himself made such great contributions in the same field. Armstrong said that when he left Kolbe early in 1870 'he was already peculiar; he afterwards, in his last years, so fixed his mind upon certain grievances as to be little short of a monomaniac'. Victor Meyer, and there can be no better judge, sums up the Kolbe paradox in a sentence: 'Zwar wurde Kolbe in seinen späteren Jahren ein unermüdliche Bekämpfer der Valenztheorie, doch hat grade er in erster Linie zur Klarlegung des Valenzbegriffs beigetragen.'

Armstrong's final word on the Kolbe–Kekulé controversy was in his review in *Nature* of Anschütz's *Life of Kekulé*:

Nothing is more certain than that most of us only take in new ideas through experience—the want must be felt before it can be satisfied. Once assimilated, an idea is expelled or modified with great difficulty. It is that that makes scientific thought, the scientific habit of mind, so difficult of attainment. Kekulé at once fell victim to Gerhardt's magic influence, when he met him in Paris. His belief in Gerhardt's system became strengthened, in London, through association with Williamson and Odling. He does not seem to have been intimate with Frankland. He appears to have been so satisfied with the superiority of Gerhardt's system, that he took little, if any, notice of Kolbe's work: I do not believe that he ever mastered the inner meaning of Kolbe's formulae. Kolbe had little use for the Gerhardt formulae, knowing that he had penetrated deeper than they carried him. I feel sure he resented the way in which he and Frankland were waved aside by Kekulé: and probably, this was the subconscious, if not conscious, primary cause of the bitterness he displayed towards him, in later life. In addition he was a linguistic purist and idealist, and was greatly annoyed by Kekulé's at times flamboyant masterful style. As I have said elsewhere, Kolbe's doctrine was ever the Pauline: 'Alles prüfen'—Prove all things! He took exception, therefore, to what he thought to be Kekulé's dogmatic, if not arrogant, declarations. Intellectually, Kekulé probably was Kolbe's superior, but not as a constructive worker. Frankland and Kolbe's synthesis of acetic acid [1846] is one of the most clear-cut achievements in the early history of the development of the doctrine of chemical structure: Kekulé seems never to have grasped its significance and the extent to which it put their work in advance of his.

My last talk with Armstrong before we reached Tangier was about the final episode in the battle of atomic weights, the Karlsruhe Conference.

In spite of the success of Gerhardt's classification among the younger chemists, the gradual realization of the combining powers of elements, and Kekulé's application of this idea in the theory of atom linking, chemistry in 1859 was still in chaos. There was no general agreement as to the basis of atomic weights or molecular magnitudes, and the various formulae proposed for acetic acid covered a page in Kekulé's *Lehrbuch*.

Kekulé saw that the acceptance of his theory of atomic linking depended on agreement about atomic weights, and in the autumn of 1859 he suggested to Weltzien, Professor Chemistry at Karlsruhe, the idea of a Chemical Conference to discuss some of the fundamental issues that were in dispute. Weltzien met Kekulé and Wurtz in Paris, and they sent a joint letter to their colleagues abroad asking their views as to the usefulness of such an international gathering. As a result an invitation went out from the organizers inviting them to a three-day conference at Karlsruhe on 3 September 1860. Over a hundred chemists came from almost every European country; Liebig, Wöhler, and Mitscherlich were absentees, but most of the active workers were present, so that the organizers must have hoped for great results. But there is a sentence in Gerhardt's treatise which proved only too true a forecast.

'What I do not understand is that when chemists meet for discussions, each speaking his own language, such discussions are always fruitless, even when the chemists are in complete agreement as to the facts—either because, without realizing it, each expresses the same facts in a language which his opponent does not understand or because they all give to formulae a significance they cannot have, that of representing molecular structure.'

An account of the proceedings of the Karlsruhe Conference will be found on pp. 185–94 of this book. In spite of its inauspious ending, its objective was ultimately achieved, thanks largely to the copies of Cannizzaro's précis of his lectures to his students on atomic and molecular weights that were given to Lothar Meyer, Mendeleev, and others. However some little time elapsed before all chemists accepted Cannizzaro's reformed atomic and molecular weights. Armstrong told me that when he began to learn chemistry, two formulae for water, HO and H_2O, were still in use!

Henry Armstrong 219

And this brings us to Armstrong's own training, which began in 1865, the year that saw the birth of the benzene ring, which was to absorb so much of Armstrong's thought and work. He 'just slid into chemistry' under Frankland at the Royal College of Chemistry, and the story is best told in his own words. After four terms,

Frankland saw, I think, that to keep me at lessons was waste of time: at all events, he wanted someone to help him and paid me the compliment of asking me to work for him. . . . We were mainly concerned in devising methods for the determination of organic impurity in sewage and of sewage-matter in drinking waters. Frankland gave me only the barest instructions and left me to do the experimental work—the experience was invaluable: through it, at the age when I was only due conventionally to attend either Oxford or Cambridge, I became a confident, independent worker.

I have always taken it as a high compliment and proof of his unselfishness that he advised me, in the summer of 1868, to go abroad and study under his old friend and companion at arms, Kolbe, then Professor at Leipzig.

[Kolbe was] one of the most thorough and typical Germans of the old school it has been my good fortune to meet, a chemist who received but scant justice even from his own countrymen—few realize the extent to which he was the founder of our modern system of constitutional formulae—because he dared to criticize and expressed himself in the biting terms of a clear and concise diction, in a pure German which no one else in those days had at his command: in fact, he took his countrymen greatly to task for their slovenly language. Of course, I was received with utmost cordiality. I well recollect how, on the afternoon of my arrival, Kolbe took me into his private laboratory and carried out with me the nitration of a quantity of phenolparasulphonate by means of sodium nitrate and sulphuric acid. He then suggested that I should take up the study of the mixture formed on sulphonating phenol. In those days we had not yet learnt even to distinguish three isomeric mono-derivatives of benzene and phenol*ortho*sulphonic acid was unrecognized. It was at this time that Kolbe began 'to slang' Kekulé over his benzene formula—I had not even heard of this before going to Germany. Such was my introduction to the 'aromatics' and the beginning of my affection for sulphonic acids. . . .

Kolbe's laboratory, in those days, afforded wonderful opportunities. About a dozen of us were doing advanced work, in preparation for the Degree—seeking independence. Each had his

Arbeit—his definite problem—in view, as his chief aim in life: we were all proud of being called on to show that we could do something. This was the distinctive feature of the German system. At most two or three had themes from the Professor—the rest were carrying out ideas of their own; the work was, therefore, varied. Whatever suggestion we made to Kolbe, he never discouraged us; his habit was to grasp the lapels of his coat, then to reply: 'Try it, try it.' We disputed with him constantly before the blackboard, often for hours together, nearly always taking exception to his theoretical views—but without his being offended. And we constantly compared notes together. Each of us, therefore, was interested in the solution of a whole series of problems.

Armstrong stayed at Leipzig till 1870, and these years with Kolbe left their mark on him. From Frankland he had learnt experimental technique, confidence in his power to attack a problem and the call of research. But Frankland was an individualist; he never founded a school, and as Armstrong said of him: 'Frankland was so thrown upon himself, he so developed the art of self-help that he never learnt to order and use others sufficiently, which is the teacher's art; he kept counsel with himself.' Frankland was a fine lecturer, but he disliked the routine of laboratory teaching.

Kolbe, on the other hand, was an outstanding teacher, and a great inspirer of research. In spite of his theoretical eccentricities students flocked to his laboratory as they had previously gone to Liebig at Giessen. Armstrong owed to Kolbe his introduction to aromatic chemistry which came to be his most continuous and active scientific interest, and from him he learnt the value of that daily intercourse with students in the laboratory. Probably, too, his own provocative attitude towards easily accepted theories was influenced by Kolbe's hard hitting polemics.

Wherever Armstrong taught—at the London Institution, at the Finsbury Technical College, or at the City and Guilds Central College—his students were trained in method, they learnt to think for themselves. The result is seen in the unceasing flow of papers that came from his laboratory for half a century, inspired by him and representing the work of men who had the privilege of working side by side with him in the laboratory.

Although those papers contain no outstanding discovery, they covered many fields, and were a great contribution to our knowledge of the orientation of aromatic derivatives, the structure of the terpenes, the nature of enzyme action, and the morphological relations of the crystals of organic compounds. Armstrong always chose problems of real significance, but perhaps his greatest contribution was the stimulus he gave to his colleagues, his students, and his friends.

His mind was constantly ranging over a much wider field than chemistry, and while that may have diminished the intensity of his work on specialized subjects, it was the secret of his personality. I rather think that nature had meant him to be a biologist, as he had the eye and instincts of a naturalist, and a great love and understanding of living things; and he had too an intuitive perception of the things that matter to human life.

Armstrong was both a prophet and a pioneer. In so many directions he saw things ahead of his time, and having seen them, like his favourite writers, Ruskin and Carlyle, he lost no opportunity of preaching the gospel in trenchant phrase—the place of science in education, methods of teaching, the effect of diet on health, the value of fresh food, especially milk, the place of science in agriculture, the need for surveys of natural resources, fuel efficiency, and conservation. His utterances on all of them had the prescience that sets men's minds stirring long before the current of thought and knowledge has made things obvious. Christ's Hospital and Rothamsted were the larger laboratories where many of his ideas bore fruit.

In 1916, Armstrong was one of those who saw most clearly that when the war was over we had to face 'a more enduring and difficult struggle in the fields of industry and commerce. The race will be neither to the swift nor the strong, so much as it will be to the intelligent and the persevering.' He saw that coal was a key factor in our national economy. 'At present our nation is without a coal conscience; get one it must without delay. In some way the public must be made to realize how absolutely coal touches our civilization at every point.' Armstrong was a persistent and forceful advocate of the need for scientific method in the use of our national coal resources, and his was one of the voices that led to the formation of the Fuel Research Board in 1917. But it took yet another war to drive home the need for fuel efficiency that Armstrong saw so clearly thirty years earlier.

Armstrong had both a very active brain and a pen that could keep pace with the rapidity of his thoughts, aided by wide reading and a responsive memory. All his occasional papers and addresses had such vigour and freshness and a character of their own. Today they are scattered and inaccessible, and a little volume of them would be a most fitting memorial to Armstrong's memory.

In his article on Sanderson of Oundle—'The Fundamental Problems of School Policy and the Cult of the Turned-up Trouser Hem',—Armstrong is at his best, tilting at convention and the opposition to new methods. He and Sanderson were both 'heretical warriors' in the cause of education. Armstrong admired the courage of Sanderson's experiments, but his realistic eye was quick to see the weakness of Sanderson's methods. For Armstrong, the teaching of science meant method and measurement—hard discipline rather than romance. Racy and full of apt allusions, the paper ends on a disillusioned note.

'The world is selfish to the backbone. Civilization is exhausting natural resources without a thought for the morrow. We are as the Gadarene swine—rushing to destruction.' What would Armstrong have said today?

I will end as I began with a quotation—the inimitable sentences in which Sir Frederick Keeble has summed up Armstrong's character:

Armstrong was one of the great personalities of his time. He belonged not to the august order of the Olympians, but was one of the Titans. He had all the spare parts of genius, but not the long patience to put them together.

Caring more for causes than for his own advancement, he had to make his way to greatness without the goad of personal ambition to prick him on; though disinterested love of truth and tireless pursuit of knowledge were always with him as good companions to cheer him on the way.

PLATE 13

H.B.H. in 1910

Henry Moseley

PLATE 15

Sir Cyril Hinshelwood's Research Laboratory in Trinity College, Oxford. Painted by Sir Cyril Hinshelwood about 1938

10

THE CONTRIBUTION OF THE COLLEGE LABORATORIES TO THE OXFORD SCHOOL OF CHEMISTRY†

FROM its foundation in 1683 to the middle of the nineteenth century, the Old Ashmolean was the official home of chemistry in Oxford. The original inscription over its doorway was *Museum Ashmoleanum Schola Scientiae Officina Chimica*. From the introduction of the Honours School of Natural Science in 1850 until the building of the Dyson Perrins Laboratory in 1915 and the Nuffield Laboratory in 1941, the College Laboratories were responsible for part of the teaching of chemistry and were the homes of notable research.

Magdalen built the first College Laboratory in 1848, followed by Balliol in 1855, with extensions in partnership with Trinity in 1879 and 1897. Next came Christ Church in 1866. Later, with the growth of the Chemistry School, Queen's and Jesus built laboratories in 1900 and 1907. In my account of their contributions and of the men who worked in them I shall deal with them in the historical order of their foundation.

THE DAUBENY LABORATORY AT MAGDALEN

The Daubeny laboratory was built at his own expense in 1848 by the polymath, Charles Giles Bridle Daubeny, praelector for the teaching of science in Magdalen from 1820 until his death in 1867, Aldrichian Professor of Chemistry from 1822 to 1854, Sherardian Professor of Botany, 1834 to 1867, and Professor of Rural Economy from 1840 to 1867. Dissatisfied with the gloomy laboratory in the basement of the Ashmolean Building in Broad Street in which he had to lecture, Daubeny, with the agreement of Magdalen, built a substantial block in the Physik Garden in

† Reprinted from *Chemistry in Britain*, November 1965.

which he lectured on chemistry until Brodie succeeded him in 1854, after which he gave Catechetical Lectures there for the School of Natural Science to members of Magdalen until his death in 1867. Edward Chapman was appointed lecturer in 1869 and continued to lecture on the physical sciences and give laboratory instruction until 1894. From 1874 to 1885 the Daubeny Laboratory under Chapman and C. J. F. Yule was the active centre of the teaching of physiology in Oxford and it was during this period that H. B. Dixon, who had been working on the rates of the explosion of gases in the Christ Church Laboratory, visited Magdalen for two years to use Yule's electric chronograph to measure the rates of these explosions. In 1883 he bought the chronograph and moved it to Balliol.

The appointment of J. J. Manley in 1888 as Daubeny Curator to succeed John Harris, Daubeny's assistant, had a decisive influence on the history of the laboratory, as Manley was a gifted experimenter and for forty years his clever fingers and observant eyes were engaged in research, while at the same time he taught science in Magdalen College School and held practical classes for undergraduates. Manley was always ready to assist his botanical colleagues with analytical data and in 1893 he began a series of investigations with V. H. Veley on the physico-chemical properties of solutions of nitric acid which continued until 1901, about which time he and D. R. Wilson gave a course of practical work in physical chemistry in the laboratory. In 1904, under the co-operative scheme introduced by H. B. Baker, Magdalen undertook to provide a course of quantitative analysis under Manley which was largely attended.

Manley's interest lay in accurate measurement and he was a true craftsman, making his own instruments. His papers on improvements in the construction and use of balances added considerably to their accuracy, leading up to his own classic paper in 1912 on the constancy of mass during chemical reactions. His generous help and advice were constantly in demand by his younger colleagues who profited by his skilful workmanship during the early decades of this century.

In 1901, with the increasing numbers reading chemistry, more space was required and two new laboratories had been added, the upper one for undergraduate teaching and the lower for research, the latter being devoted mainly to physical chemistry. It was here that Sidgwick began his Oxford research after his return from Tübingen to Lincoln in 1901, as he was teaching both at Lincoln and Magdalen until T. S. Moore was appointed by Magdalen in 1905. Among Sidgwick's pupils who worked with him in the Daubeny Laboratory were H. T. Tizard, A. C. D. Rivett, and B. H. Wilsdon, on subjects ranging from the colour of solutions of copper salts to the rate of hydration of acid anhydrides and the study of solutions of aniline salts. Meanwhile Moore and T. F. Winmill were carrying out their important determination of the true ionization constants of ammonia and various amines, which prompted their suggestion of a hydrogen bond to explain the weakness of all but the quaternary amines. Donald Somervell (later Home Secretary and Lord of Appeal) was another of Moore's research pupils.

Moore left Magdalen in 1914 and after the war Manley continued to give his practical classes until the Daubeny Laboratory ceased to be used for teaching chemistry in 1923. Manley, who had been elected to a Research Fellowship at Magdalen in 1917, continued his research into the refinement of platinum thermometry and the construction of balances until he left Oxford in 1929.

THE BALLIOL AND TRINITY LABORATORY

The statute of 1850 forced the Colleges to consider how they would provide tuition for the School of Natural Science. Balliol, under the influence of Jowett, decided in 1851 to build a chemical laboratory in the cellars of the Salvin Building, known later as No. 16, which was to perpetuate the subterranean tradition of chemistry in Oxford for ninety years. Henry Smith, the witty mathematical tutor and Ireland Scholar, was sent for a few months to learn some chemistry under Hofmann at the Royal College of Chemistry in Oxford Street and he also worked with Story-Maskelyne in the cellars of the Old Ashmolean.

Augustus Vernon Harcourt, one of Henry Smith's first two pupils, was destined to be the key figure in the development of teaching and research in college laboratories, as five of his pupils were elected into the Royal Society and the successive generations of their pupils provided nearly all the heads of College Laboratories, besides contributing many other distinguished teachers of chemistry. I was made to realize the strength of the Harcourt tradition when my tutor, Sir John Conroy, saw me commit one of the minor crimes of the laboratory. He looked at me sadly and all he said was: 'Harcourt would have told Dixon, Dixon would have told Baker and Baker would have told you'. When I was telling that story, someone broke in with 'Yes, and I suppose you would have told Hinshelwood'. 'That', I said, 'was quite unnecessary'.

When some phosphorus caught fire in the cellar laboratory Harcourt's companion was going to pour water on it. Henry Smith, however, extinguished the blaze with sand, remarking in his soft tones: *'Pulveris exigui jactu compressa quiescit.'*

When B. C. Brodie (later Sir Benjamin) succeeded Daubeny as Aldrichian Professor in 1854, his College put the Balliol cellars at his disposal until the Abbot's Kitchen in the New Museum was built. It was there that Brodie began working on the peroxides of the acid radicals, on the oxidation of graphite and on a systematic study of the alkaline peroxides. Harcourt was first his pupil and later his assistant, and when Brodie moved to the Museum, Harcourt, still an undergraduate, went as his lecture assistant. In 1859 he was elected to a Lee's Readership at Christ Church after getting a First Class in Schools.

There is no record of events in the Balliol Laboratory after Brodie left until 1879 when Balliol and Trinity joined forces in the teaching of chemistry in a new laboratory built by Balliol on a strip of land between the New Hall and the Trinity boundary, with a lecture room behind the Senior Common Room in which Ruskin once lectured on glaciers. Harcourt's pupil H. B. Dixon was put in charge of the laboratory, holding lectureships in both Colleges. A door was opened in the wall between the Colleges to give Trinity men access to the laboratory and in 1897 the frontier was breached again to connect the old laboratory with a new section built in Trinity. Dixon continued his investigation of the velocity of gaseous explosions in No. 16, moving his apparatus to the Hall when he wanted to study the progress of the explosive wave over longer distances. Meanwhile his pupil, H. B. Baker, was making a systematic study of Dixon's discovery of the effect of dryness on chemical reactions. On one occasion Dixon and Roscoe, returning from

Contribution of the College Laboratories

a holiday, were surprised to find Baker distilling phosphorus in dry oxygen. Among Dixon's other pupils were Spencer Pickering, J. E. Marsh, Viriamu Jones, and Daniel Hall.

In 1887, when Dixon went to Owens College, Manchester, Sir John Conroy, another of Harcourt's pupils, was appointed in his place. He had been the science tutor at Keble since 1880, where he had built a small laboratory which was used to teach Preliminary Physics until 1904. Conroy brought some of the benches with him which were installed in yet another Balliol cellar under the Hall and here he did his research. His colleague D. H. Nagel was appointed Millard Lecturer by Trinity in 1888 and a Fellow in 1890. Conroy was basically a physicist and his research was mainly on optics. He was often abroad on account of his delicate health and he died in Rome in 1900. His bequest of apparatus and books was a great help when Balliol and Trinity undertook to provide a practical course in physical chemistry in 1904.

In Conroy's day No. 16 was often in use. Lord Berkeley had worked for a time in Christ Church after coming to Foxcombe in 1893 and later he moved to No. 16 to investigate with E. G. J. Hartley the electrolysis of glass following on some observations of Roberts Austen. This aroused his interest in semi-permeable membranes and led to his classic work on osmotic pressure which began in No. 16 before he had built his laboratory at Foxcombe in 1898.

In 1899 Soddy started his research in No. 16 and christened it the Brodie Laboratory. I asked him what he was doing. 'I am making one half of Marsh's formula of camphoric acid, and Harold Carpenter (another Merton man) is making the other half in Manchester. It is such a stable substance that when we put them together they will combine with a click.' At the end of the year, Soddy packed his bag and joined Rutherford in McGill and Carpenter turned to metallurgy.

I succeeded Conroy in 1901 and with the help of my pupils did research on phase-rule problems and spontaneous crystallization, and later on the electrochemistry of non-aqueous solvents. Although Nagel had never undertaken any research, his wide knowledge and his ingenuity in suggesting improvements in experimental techniques were an immense help to many of us. I remember Moseley coming from Manchester to consult his old tutor when Nagel selected the cleavage fragment of potassium ferrocyanide which Moseley used in his classic investigation of X-ray spectra. Sidgwick once said of him: 'Never ask Nagel whether you should undertake a particular research, but when you have made up your mind to do so, his advice will be invaluable.'

Many undergraduates from other colleges came to the laboratory for the course in physical chemistry which was largely planned by Nagel. As the numbers increased more cellars were invaded. After the war, when Hinshelwood succeeded Nagel, the buildings in Dolphin Yard, one after another, found new and unwonted uses to provide him with space for his work on the kinetics of gas reactions, and also for E. J. Bowen's photochemical investivations. The first edition of Hinshelwood's *Kinetics of Chemical Change in Gaseous Systems* was a milestone in the literature.

In 1930 when I left Balliol to go to Euston I could look back on a series of pupils who were to play their part in chemistry in Oxford and elsewhere; Thomas Merton, R. B. Bourdillon, Humphrey Raikes, E. J. Bowen, Cyril Hinshelwood, Sydney Barratt, J. J. Wolfenden, Oliver Gatty, W. F. K. Wynne-Jones, R. P. Bell, J. St. L. Philpot, and A. G. Ogston. Oliver Gatty succeeded me for a short time, followed by Ronnie Bell, whose distinguished work on catalysis added another stream to the long list of researches from the laboratory, and to this Hinshelwood added yet again when he began to investigate the adaptability of living cells in 1938.

In 1937, when Hinshelwood succeeded Soddy as Dr. Lee's Professor, the practical teaching of physical chemistry was still carried out in the Balliol and Trinity Laboratory and also in Jesus under an arrangement with the University. This continued until 1941 when Hinshelwood moved to Lord Nuffield's fine laboratory in the Parks, and the Balliol cellars and the Trinity outhouses returned to their normal functions, after long and honourable service to chemistry.

THE CHRIST CHURCH LABORATORY

After Vernon Harcourt's election to a Lee's Readership in 1859 he continued to work at the Museum until 1863, when the two-storeyed building erected in 1767 under Dr. Lee's Bequest as a

School of Anatomy on a site south of the Dining Hall was converted to a laboratory for the teaching of inorganic chemistry and elementary physics. The anatomy collection had been moved to the University Museum in 1859. Harcourt's main research was on the rate of chemical change, in which work he was associated with William Esson, later professor of geometry, who was responsible for the mathematical interpretation of Harcourt's results. The reaction between oxalic acid and permanganate proved to be complicated by the secondary action of manganese sulphate, but in the reaction between hydrogen peroxide and hydrogen iodide they found one suited to their purpose and their results together with those of Berthelot and Guldberg established on a quantitative basis the law of mass action. In studying the effect of temperature on the rate of reaction they arrived at a zero of chemical action in close agreement with the absolute zero based on physical data. Harcourt was also interested in applied chemistry and as a Gas Referee he worked on the removal of sulphur from coal gas; and he designed the Pentane Lamp which was used as the standard of illumination. Later he became interested in the safe administration of chloroform as an anaesthetic and his inhaler was largely used. Harcourt was a most painstaking teacher and among his pupils were Sir John Conroy, H. B. Dixon, P. Elford, F. D. Chattaway, D. L. Chapman, N. V. Sidgwick, E. G. J. Hartley, and Andrea Angel, who all played an active part in Oxford's chemistry.

From 1879 to 1888 V. H. Veley (University College) began his research under Harcourt's eye at Christ Church and his early papers on rates of reaction and on the lime process for the purification of coal gas bear evidence of Harcourt's influence.

H. B. Baker succeeded Harcourt in 1903, when a third storey was added to the laboratory. Baker at once proposed that each college laboratory should undertake responsibility for one branch of practical work and thus gain the experience and techniques needed for a research group. Under his scheme Balliol and Trinity provided a course in physical chemistry, later supplemented by Jesus, while Christ Church undertook inorganic chemistry and Magdalen quantitative analysis. Organic chemistry was taught at the Museum and later at

Queen's. There is no doubt that Baker's initiative led to the large increase in published papers from Oxford after 1906.

Baker resumed his work on the effects of the intensive drying of substances by phosphorus pentoxide, begun with Dixon at Balliol and continued at Dulwich. Substances like ammonium chloride showed no dissociation when heated if they were Baker-dry. He began there the work on the abnormal boiling points of liquids when they had been intensively dried. Angel succeeded Baker when he left for Imperial College in 1913 and after his death in the Silvertown explosion in 1916 there was an interregnum until A. S. Russell was appointed in 1920. Russell continued the long tradition of research in Christ Church by his investigation of the chemistry of protoactinium and other problems in radioactivity. Later he worked on intermetallic compounds in the medium of mercury and supervised the work of Part II students both for Christ Church and other colleges. From 1920 he lectured at Christ Church on inorganic chemistry to undergraduates reading engineering.

The 'utilitarian' top storey, erected in 1903, proved an eyesore when the Memorial Garden of Christ Church was planted in 1930, and was removed. After 1954 the building was converted for use as a picture gallery and a museum on one floor with a refectory on the other.

THE QUEEN'S LABORATORY

In 1900, G. B. Cronshaw opened a small laboratory at Queen's which, with A. F. Walden's assistance, was used mainly for preliminary teaching until in 1907 Chattaway returned to Oxford after sixteen years spent at St Bartholomew's Hospital or abroad. He extended the work at Queen's to include organic chemistry and the laboratory was soon enlarged for this purpose, Chattaway becoming its head. He quickly built up an active and most productive school of research and, until Perkin arrived and the Dyson Perrins laboratory was built, Queen's was the centre of organic chemistry in Oxford. During Chattaway's tenure 124 papers appeared from the laboratory, 110 of which bore the names of his collaborators. He had a genius for making difficult reactions work and for overcoming dangers that others had encountered. In fact they seemed to attract him. He was an admirable but exacting teacher and many men owed to

Contribution of the College Laboratories

him their skill in organic techniques. In addition to his research pupils he had an undergraduate course in organic chemistry which was largely attended.

Chattaway's publications at Oxford covered a wide field. Many of them dealt with the action of the halogens on various organic compounds and the properties of the derivatives that were obtained, including the alkyl hypochlorites. Other investigations dealt with the reactions of chloral, the configurations of the trithioacetaldehydes, the reaction between glycerol and oxalic acid and a safe method of making the potassium salt of aci-trinitromethane. All are examples of Chattaway's experimental ingenuity.

THE LEOLINE JENKINS LABORATORY AT JESUS

In 1907 a well-equipped laboratory was built by Jesus, and D. L. Chapman, who had spent ten years with H. B. Dixon at Manchester after leaving Christ Church, was appointed its head, a position he held until the laboratory was closed in 1944. It was there that Chapman, with the help of his able wife and a long line of pupils, carried out research on chemical kinetics including the homogeneous decomposition of ozone, the reaction between chlorine and carbon monoxide, and the influence of metals on the combination of hydrogen and oxygen. Chapman's major contribution was his long series of investigations into the mechanism of the photochemical reaction between hydrogen and chlorine, and its inhibition by the presence of small quantities of various impurities, by which he finally established the chain mechanism of the reaction. During the 1914–18 war, Chapman was a member of the Inventions Committee and many of the inventions were tested in his laboratory. During the 1939–45 war he worked with a small team on diffusion processes for the Tube Alloy programme.

Chapman was an excellent teacher and lecturer and under his leadership the Jesus Laboratory was for over thirty years an active centre of teaching and research in physical chemistry and made a notable contribution to the strength of the Oxford Chemistry School.

The most fitting ending to this story is Vernon Harcourt's final sentence in his Address on the occasion of the Jubilee of the Museum in 1910: 'In the Museum, sometimes one study advances more rapidly and attracts more pupils, sometimes another . . . but when "the tired waves vainly breaking, seem here no painful inch to gain", the College laboratories with young and eager teachers are prepared to supply any deficiency that may occur.' This was true when those words were spoken and so the contribution of the college laboratories to the Oxford School of Chemistry should not be forgotten.

INDEX

Acids, 32, 36f., 79, 109, 118, 147f., 176, 205–6, 208, 211–13, 227, conductivity of, 175. Acetic, 84, 150f., 186, 201, 217; boracic, 108; carbonic, 37, 41f., 45, 68, 72, 82, 211; fluoric, 108, 115–16; hydrochloric, 12; lactic, 213; malic, 213; muriatic, 96, 108, 111, 115, 164; nitric, 7, 9, 12f., 36f., 45, 65, 69, 118, 224; oxalic, 74; oxymariatic, 110–11, 115; phosphoric, 37, 110; polybasic, 150; salicylic, 213; sulphuric, 12, 42, 68, 72, 148, 164; tartaric, 85, 213
Agricultural chemistry, 112
Air, 6–7, 10–12, 31ff., 36–7, 40, 64, 72, 101; air pump, 2, 94; apparatus for experiments, 44; 'dephlogisticated', 11–12, 34; inflammable, 42–4; nitrous, 8–11, 14
Alchemy, 6–7, 27, 46, 108
Alcohol, 82f., 148, 205, 210; secondary and tertiary alcohols, 212
Alkalis, 26, 28ff., 47, 107ff.
Alternating current, 174
Ammonia, 8, 12f., 29, 45, 99f., 108, 113, 225
Ammonium amalgam, 137f.
Ampère, A. M., 202; and Davy, 117; electrochemistry, 158, 186, 191; fluorine, 115f.; molecular theory, 76–7, 146
Anaesthetics, 9, 229
Analysis, qualitative and quantitative, 143, 149, 176, 229
Anderson, Sergeant, Faraday's laboratory assistant, 155, 166

Angel, Andrea, 229f.
Animal chemistry, 138, 148; *see also* Physiological chemistry
Animals and birds, experiments with, 99; birds, 48; guinea-pigs, 41–2, 48; mice, 6–7, 10–12
Annales de Chimie, 76
Annals of Philosophy, 75
Annual Reports, see under Berzelius
Anschutz, R., *August Kekulé*, 188, 217
Apreece, Mrs., *see* Davy, Lady
Arago, D. F. J., 158
Armstrong, H. E.: biographical information, 196, 218–22; characteristics and opinions, 195–9, 206, 219–22; *and* inorganic chemistry, 197; with reference to Frankland, Kolbe, and Kekulé, 206–18; to Gerhardt and Laurent, 199–206, 213–18. *See also* Frankland Memorial Oration
Arrhenius, S. A., 173, 175ff., 179, 181
Arsenic, 144
Atomic chemistry and theory, 59, 74–89, 97, 140–3, 149, 186. *See also under* Berzelius; Dalton; Newtonian atom
Atomic heats, 79, 84; linkage, 212f., 218; structure, 83, 86; symbols, 190f., weights, 73, 75, 78–80, 83ff., 92, 142–6, 150, 152, 186–7, 189–93, 201f., 211, 214, 217
Avogadro, A., 76f., 81, 83f., 146, 175, 186, 191f.
Azote, *see* Nitrogen

234 Index

Babbage, C., 124
Bacon, Francis, 20
Baker, H. B., 134, 196–7, 224, 226, 229f.
Bakerian lectures, 74, 105–9, 129, 161, 165
Balard, A. J., 82, 204
Balliol College, Oxford, 223–30
Banks, Sir Joseph, P.R.S., 103f., 117, 122, 124, 126
Barium, 108
Barratt, Sydney, 228
Beccaria, G. B., 31
Beddoes, Thomas, 96f., 99 and n, 101f.
Beddoes, Mrs., 97
Beetz, W., 174
Bell, R. P., 228
Bentham, Jeremy, 1
Benzene, 85, 154, 216, 219
Bergman, Torbern, 134f.
Berkeley, Lord, *and* electrolysis of glass, 227
Bernard, Sir Thomas, 102
Berthelot, P. E. M., 175, 188, 229
Berthollet, C. L., 45, 48, 59, 64, 72, 75f., 89, 93, 95, 136, 138
Berzelius, J. J., 55, 59, 83, 89, 106f., 111, 208f. Achievements, 78–81, 92, 108, 136, 141–9, 151–2; biographical information, 135, 137, 146, 151, 155; characteristics, 58, 78–9, 135–40, 143, 149, 151–2; authority challenged, 149–51, 199, 201–2; early research, 137–8; papers on diverse subjects, 149
Annual Reports, 8off., 132, 146–8, 151; *Essay on the Theory of Chemical Proportions*, 143–5; *Lehrbuch* and *Textbook*, 72, 81, 85–6, 143, 148, 151, 199; *Organic Chemistry*, 214
And Dalton, 72f., 76, 139–40; Davy, 108f., 113f., 127, 132f.; Kolbe, 206–7; Mitscherlich, 144–5
And atomic chemistry and theory, 72ff., 76–82, 84f., 140–3, 145f., 149f., 152, 187, 190ff., 202, 210; electrochemistry, 186; laboratory devices, 143
See also *Jahresbericht*
Berzelius Museum, 134
Biot, J. B., *and* optical rotation, 86, *Physics*, 144
Bjerrum, N., 177f., 182
Black, Joseph, 3, 5, 16, 19f., 28ff., 40, 55, 97, 132
Blagden, Sir Charles, Secretary of the Royal Society, 43
Board of Longitude, 125, 156
Board of Ordnance, 156
Bohr, N., *and* atomic structure, 87, 177
Borlase, Bingham, surgeon, *and* Davy, 94, 97, 132
Boron, 109, 209
Boulton, Matthew, 14
Bourdillon, R. B., 228
Boussingault, J. B. J. D., *and* Karlsruhe conference, 189
Boyle, Robert, 4f., 20, 25f., 30, 44
Brande, W. T., 122
Bréhant, 100
Brewster, David, 78
British Museum, 125f.
Brodie, Sir B., 224, 226
Brouncker, William, Viscount, P.R.S., 4
Bucholz, C. F., *and* quantitative analysis, 59, 138
Bucquet, J. B. M., *and* inflammable air, 42
Buddle, John, quoted on safety lamp, 120–1
Bunsen, R. W. von, 188, 207ff.
Butane, 208
Butlerow, A. M., *and* tertiary alcohols, 212
Butylene, 147, 154

Cacodyl, 208ff.
Cadet-Gassicourt, L. C., *and* Lavoisier, 26, 33
Calcination, 29–34, 36
Calorimeter, 40–2, 123

Index

Cannizzaro, S., biographical information, 186, lectures, 186–7, 218, *Sunto*, 185, 192; and Karlsruhe conference, 185, 187–93, 218; and atomic theory, 84f., 89, 146; and vapour densities, 83
Carbon, 45, 68, 70, 80, 83, 85, 212, 215, compounds, 185, 216
Carbon dioxide, 3f., 26, 29f., 40, 48–9, 96, 112, 118
Carbonates, 29
Carbonic oxide, 15
Carbonyl, 212
Carpenter, Harold, 227
Catalysis, 81, 148, 152, 176, 228
Cavendish, Henry, 5, 7, 13f., 16, 19f., 23, 32, 43, 45, 55, 68f., 136, 159
Celsius, A., and magnetic theory, 62
Chapman, D. L., 229, 231
Chapman, E., 224
Chaptal, J. A. C., 48
Chattaway, F. D., 229ff.
Chemical affinity, 106, 158, 169, 177, 201
 combination, 73–6, 113, 118
 equations, 47, 82–3, 135, 214
 formulae, 92, 141–5, 149f., 152, 186, 201–7, 209–12, 214ff.
 notation, 189f.
 proportions, 139, 143
 reaction, 226, 229ff.
 reagents, 211f.
 symbols, 142, 189f.
 See also Nomenclature
Chevreul, M. E., 80
Chlorine, 81, 111, 113f., 142, 149–50, 153
Chlornaphthalenes, 81, 149f.
Christ Church, Oxford, 223f., 226–30
Cigna, G. F., and phosphorus, 26f.
City Philosophical Society, 158
Claubry, Gautier de, 77, 140
Clausius, R. J. E., and Ohm's Law, 175
Clerk-Maxwell, J., 172
Clerke, Dr. F.R.S., 4

Coleridge, S. T., and Davy, 97–9, 101
Colour blindness, 62
Combustion, 5, 7–11, 26–33, 36f., 39–45, 47, 123
Comptes Rendus mensuels des Travaux Chimiques, 202
Condillac, E.-B., 20, *Logic*, 46
Conductivity of solutions, 173–83
Conroy, Sir J., 226f., 229
Copley medal, 104
Copper, 128–9
Cottle, Joseph, Bristol bookseller, 98
Coulomb forces, 178, 181
Couper, A. S., and organic chemistry, 85
Courtois, B., and iodine, 117
Crawford, A., 40
Creed, John, F.R.S., 4
Cronshaw, G. B., 230
Crosthwaite, friend of Dalton, 61
Crystallography, 76, 85–6, 144, 181, 197–8, 200, 204, 227
Culinary chemistry, 103
Cuvier, G., 117

Dalton, John:
 Achievements, 59, 71, 78, 87–9, 140, 144; biographical information and characteristics, 58–62, 70–3, 78, 87–9.
 Elements of English Grammar, 64; *Essay on the Mind*, 64; *Meteorological Observations and Essays*, 60–1; *New System of Chemical Philosophy*, 58, 71–3, 75, 77, 87, 139
 And atomic theory, 59f., 62, 66–74, 78f., 86–9, 139, 152; and meteorology, 58, 60–3, 72f., 87 Other references, 20, 55, 83, 136
 See also under Berzelius; Manchester, Literary and Philosophical Society
Dance, Mr., and Royal Institution, 116
Daniell, J. F., 171, 173

Darwin, Erasmus, 15
Daubeny, C. G. B., 223, 226
David, J. L., 23
Davy, Edmund, 93
Davy, Sir Humphry:
    Achievements, 96, 102, 104, 106f., 110–1, 126, 128–9, 132–3; biographical information, 92–9, 101–6, 108, 114–19, 124, 130–2, F.R.S., 106, P.R.S., 124–9; characteristics, 58, 92–3, 96–104, 109ff., 113, 116–20, 127, 132, 161; early experiments, 94–6
    Elements of Chemical Philosophy, 113–14; Lectures, 111–12, 114, 116, 153, 165, *see also* Bakerian lectures; *Salmonia*, 130–1
    And agriculture, 112, 132; atomic theory, 74f., 88, 97, 113, 157; corrosion of sheathing of ships, 128–9; electricity and magnetism, 127–9, 140; fishing, 130–1; geology, 111–12, 132; 'laughing gas', 9; safety lamp, 92, 119–23, 132, 154; Vesuvius, 124
    Other references, 5, 16, 47, 55, 59, 64, 67–70, 72, 78, 81, 136, 142
    See also under Berzelius; Faraday
Davy, John, 93f., 111, 121, 130
Davy, Lady, 114, 116, 121, 131
Davy Medal, 122
Debye, P., 173, 178–81, 183
Decimal system, 48
Deluc, J. A., 40, 61, 72
Desmarest, N., *and* latent heat, 40
Deville, H. E. Ste-Claire, *and* nitric anhydride, 205
Diamonds, 26, 37, 118, 164
Dimorphism, 145
Dixon, H. B., 224, 226f., 229ff.
Dualism, 141–2, 149f., 186, 201ff, 206, 210, 214
Dulong, P. L., 136, *and* Petit, A. T., law of, 145, 187, 192
Dumas, J. B. A., 84f., 145, 149, 185, 190–3, 200, 207; *Philosophie Chimique*, 81

Dunkin, Robert, *and* Davy's early experiments, 94
Du Pont, Irénée, *and* gunpowder, 35,
Duppa, B. F., *and* organic chemistry, 213

Edgeworth, Maria, with reference to Davy, 104–5, 126–7, 130
Electricity, 2f., 13, 127–9, 153f., 158–70, 177
Electrochemistry, 81, 107, 113, 129, 132, 137ff., 140–1, 149, 157, 201f., 227. *See also under* Faraday
Electrodes, 166, 173–4
Electrolysis, 102, 105–6, 159–69, 173, 184, 208, 210
Electrolytes, 173, 175–8, 180–3
Electrolytic conduction, 183; electrolytic dissociation, 173–84
Electromagnetism, 158–9, 172
Electron, 171
Elford, P., 229
Encyclopaedia Britannica, 197
Engler, C., *and* Karlsruhe conference, 188
Erdmann, O. L., *and* Karlsruhe conference, 191
Esson, William, 229
Ether, 32, 70, 83, 205, 210
Ethyl, 205, 208
Ethylene, 147
Evaporation, 61, 63f.

Falkenhagen, H., *and* conductivities at high frequencies, 180
Faraday, Michael:
    Achievements, 135, 153n., 154, 156–7, 169, 170–2; biographical information, 116, 153, 156, 158; characteristics, 138, 154–8, 172, 174; early experiments, 153–5
    Chemical Manipulation, 48, 143, 155; Notebooks, 156, 162
    And advancement of chemistry, 184; atomic theory, 157, 171; benzene, 154; butylene, 147, 154; Davy, 97, 116, 124, 128, 132, 153–4, 161; electrochemistry, 129, 153n., 158–74, 177

Index

Faraday Lectures, 89, 171, 177
Fermentation, 47
Fertilizers, 112
Figurovsky, Academician N. A., 193
Fluorine, 115f., 167
Fourcroy, A. F. de, 46
France:
    Academy of Sciences, 21-2, 25, 27f., 30, 36, 39f., 43f., 46, 49f., 52f., 86f.; agriculture, 37-9; French Revolution, 51-4; geological map, 20-1; Physiocrats, 38
Frankland, Sir Edward, 185, 199, 207-13, 215, 217, 219f.; *and* acids, 213; valency, 209f., 216; zinc methyl, 208-9, 212. *See also under* Kolbe
Frankland Memorial Oration, 196, 206
Franklin, Benjamin, 2, 8, 23f., 52
Fréchines, France, Lavoisier's farm at, 38-9
Friedel, C., *and* secondary alcohols, 212

Gahn, J. G., 134, 137
Galton, Samuel, 15, his daughter's description of Priestley, 16
Galvani, L., 138
Galvanism, 102, 104, 106, 137f.
Galvanometer, 158f. and n.
Garnett, Dr. T., 62, 102f.
Gases, 4-16, 28-30, 32-5, 43-5, 64-6, 71ff., 142; gas-analysis, 7, 70, 207f.; densities, 13ff.; diffusion, 14, 70; elasticity, 4-5; explosion, 224; reaction, 228
Gassiot, J. P., 155
Gatty, Oliver, 228
Gaudin, M. A., 191; *and* molecular theory, 77; vapour densities, 83
Gay-Lussac, J.-L., 59, 75-7, 107-10, 117, 136, 142, 145, 186
Geikie, Sir A., with reference to Lavoisier, 21
Gentlemen's Diaries, 60, 71
Geology, 20-1, 197-8, 207

Gerhardt, C.:
    Biographical information and characteristics, 84, 200-1, 204, 214-15; *Comptes Rendus*, 82; *Introduction to Chemistry by the Unitary Method*, 199, 203; *Précis de Chimie Organique*, 199, 202-3; *Textbook and chemical equation*, 83; *Traité*, 199
    And Berzelius' *Organic Chemistry*, 214; international conferences, 218; organic chemistry, 199-207; rational formulae, 207; theory of types, 206, 212-15; vapour densities, 83.
    Other references, 77, 81, 150, 185, 187, 190f. *See also under* Armstrong; Kekulé; Laurent
Geuther, J. A., *and* acetoacetic ester, 213
Ghosh, J. C., Debye *and* erroneous theory of, 178
Gilbert, Davies, P.R.S., 130, *and* Davy, 97
Gillray, James, cartoon depicting Davy, 103
Gingembre, P., *and* inflammable air, 42
Girtanner's *Elements of Antiphlogistic Chemistry*, 135
Glass, 156 227
Gmelin, L., *Handbook*, 80, 199; *and* atomic weights, 201f., 210, 214
Goddard, Jonathan, F.R.S., 4
Gordon, *and* Faraday, 154
Gough, John, 59-60, 71
Graham, T., *and* polybasic acids, 150
Grape juice, 82, 148
Greenwich Observatory, 125
Gresham College, 4
Groth, P. H., of Munich, 197
Grotthus chain, 174
Grove, Sir W. R., 155
Guettard, J-E., geologist, 20f.
Guldberg and Waage's Law of Mass Action, 176f., 229
Gunpowder, 34-5

Guyton de Morveau, L. B., Baron, 46

Hadley, G., 61
Hales, Stephen, 5, 7, 20, 27f., *Vegetable Staticks*, 4
Hall, Daniel, 227
Haller, A. von, 41
Halley, E., 62
Harcourt, A. V., 225–9, 231
Harris, J., Curator of Daubeny Laboratory, 224
Hartley, E. G. J., 227, 229
Hatchett, C., 122
Haüy, R. J., 54, 82, 144, 147, Law of Rational Indices, 76
Heat, 40–2, 71f., 79, 94–5; conductivity of, 13, 63. *See also* Thermochemistry
Helmholtz, H. L. F. von, 171f., 177
Henry, William, 65, 71, 88
Herschel, Sir John, 124
Higgins, B. and W., 88
Hinshelwood, C., 226, *Kinetics of Chemical Change in Gaseous Systems*, 228
Hisinger, W., Berzelius *and*, 137, 143
Hittorf, J. W., *and* conductivity of solutions, 171, 173, 175
Hoff, J. H., van 't, 85, 91, 176f., 206
Hofmann, A. W. von, *and* amines, 204, 225
Hooke, Robert, 4f.
Hoyle, Thomas, friend of Dalton, 63
Hückel, E., 173, 178–81, 183
Humans, experiments on, 49–50
Humboldt, Baron A. von, 100
Hunt, Robert, F.R.S., biographer of Davy, 94
Huxley, T. H., 3
Hydrodynamics, 174
Hydrogen, 13, 15, 32, 44f., 73, 75, 79, 107–11, 113, 118, 149–50, 213f., *and* electrical charges, 177
Hydrogen chloride, 8
Hydrogen peroxide, 148
Hydrometry, 25

Hydroxides, 29

Industry, chemistry *and*, 93, 104, 156, 229
International chemical conference (1860), *see* Karlsruhe
Inorganic chemistry, 79, 142, 144, 150, 191, 197, 229f.
Iodine, 117
Ionic association, 181ff.
Ionic atmosphere, 178–81
Ionic dissociation, 197
Ionic theory, 173, 175–8
Ionization, 176f., 181, 225
Ions, 173–84, Law of Independent Mobility of, 174
Isomer, isomeric, isomerism, 81, 148, 152, 216
Isomorphism, 79, 81f., 84, 200; discovery of law of, 144–5, 147

Jahresbericht, 202, 207
Jesus College, Oxford, 223, 229, 231
Johnson, Lancaster surgeon, *and* training of druggists' apprentices, 207
Johnson, Dr. Samuel, with reference to Priestley, 1
Jones, Viriamu, 227
Joule, James Prescott, 87
Journal de Physique, 117
Journal Polytype, 45
Jowett, Benjamin, 225
Jussieu, B. de, botanist, 20

Karlsruhe, international chemical conference (1860) at, 84f., 185–94, 217–18, aim, 187–8
Keble College, Oxford, 227
Keeble, Sir Frederick, quoted with reference to Armstrong, 222
Kekulé, A.:
    Biographical information, 215; *Lehrbuch* and Textbook, 84, 199, 215f., *and* Gerhardt, 215ff.; *and* Karlsruhe conference, 84, 186–92, 218
    And acetic acid formulae, 151,

Kekulé, A.—*cont.*
186; aromatic compounds, 216; marsh type gas, 215; molecule structure, 216; organic chemistry, 85, 185, 213, 216
See also under Kolbe
Kirwan, Richard, 23, 48
Klaproth, M. H., 59, 138, 145
Kohlrausch, F., 173–5, 177, 182, with reference to Faraday, 172
Kolbe, H., biographical information, 206–8, 210, 217, 219–20; *Textbook of Organic Chemistry*, 199, 210; *and* Frankland, 211; *and* Kekulé, 216–17, 219; *and* organic chemistry, 208, 211–13; structural chemistry, 202, 206, 217
Other references, 150, 185, 209
Kopp, H. F. M., 185, 189f., 192

Laar, J. J. van, 177
Laboratories: Brodie, 227; Daubeny, 223–5; Dyson Perrins, 222, 230; Foxcombe, 227; Leoline Jenkins, 231; Nuffield, 223, 228; Old Ashmolean, 223, 225
La Caille, N. L. de, 20
Ladies' Diaries, 60, 71
Laplace, P. S., Marquis de, 40–3, 45, 48, 54, 199
Laurent, A.:
Biographical information, 82, 185, 200, 204, 214; *and* Gerhardt, 81f., 203–5, 215
And chemical equation, 83; chlorine, 149f., 200; molecular theory, 77; organic chemistry, 199f.; vapour densities, 83
Other references, 187, 190, 210
Lavergne, L. de, 51
Lavoisier, A. L.:
Achievements, 19, 27–8, 30, 37, 39, 45–6, 51, 54–6, 58, 92, 105, 107, 135; biographical information and characteristics, 19–24, 34–5, 37–9, 48, 50–6, 58, 132, 136
Chemical experiments, 25–37, 39–46, 48; *Opuscules physiques et chimiques*, 28, 30; *Territorial Wealth of France*, 51–2; *Traité Élémentaire de Chimie*, 23, 44, 46–8, 55, 72, 94f.
And chemical equation, 82; education, 53; farming, 37–9; the Ferme, 22–4, 39, 52, 54; French Revolution, 51–4; geology, 20–1; gunpowder, 34–5, 39; lighting, 21–2; meteorology, 20, 58; metric system, 53–4; physiology, 48–51
Other references, 5, 7, 9, 11, 15f., 64, 67ff., 77, 79, 81, 93, 96, 100, 110f., 140f., 152, 199, 201
Lavoisier, Madame, 23–4, 48f., 54
Le Bel, J. A., 85
Le Chatelier's Theorem, 41
Leyden battery and jars, 2, 127, 140, 159, 167
Liebig, J.: *Dictionary of Chemistry*, 210; *Organic Chemistry*, 80; Textbook, 83, 199f.
And Dalton's atomic theory, 88f.; melanine, 204; polybasic acids, 150
Other references, 136, 148f., 185, 188, 200–1, 209, 211, 215, 220
Light, 96–6, 157, 172
Lighting, 21–2, 156
Lincoln College, Oxford, 225
Linnaeus, C., 134f.
Liquids, expansion, 66; vapour pressure, 64, 72
Locke, John, 20, 46
Lockhart, J. G., with reference to Davy, 130
Lodge, Sir Oliver, 3, 174
Lubbock, Sir John, 'Hundred Best Books', 198
Luca, S. de, 186

Macquer, P. J., 26, 55
Magdalen College, Oxford, 223–5, 229
Magnetic Theory, 62
Magnetism, 157f., 172; diamagnetism, 172

240 Index

Manchester Academy, 60; Literary and Philosophical Society, 62, 87, *and* Dalton's scientific papers, 62–5, 68, 70; Owens College, 209, 227
Manley, J. J., 224f.
Marcet, A. J. G., 142; letter to Berzelius with reference to Davy, 127
Marsh, J. E., 227
Masefield, John, poet laureate, quoted, 135, 195–6
Mathematical physics, *and* ionic theory, 177
Matter, chemical properties of, 39
Mayow, John, F.R.S., 4–5, 7, 20, 28
Meldrum, A. N., 26
Mendeleev, D. I., 84f., 188f., 193, 218
Mercury, 8–11, 33–4, 36, 81, 209
Merton, Thomas, 228
Metabolism, 48–51
Metallurgy, 105
Metals, 28–33, 45, 79, 108ff., 113
  See also under names of separate metals
Meteorology, *see under* Dalton; Lavoisier
Meusnier, J. B. M., 44
Meyer, Lothar, 185, 188ff., 218; *Moderne Theorien der Chemie*, 85, 193
Meyer, Victor, *and* nitromethane, 213; with reference to Kolbe, 217
Miers, Henry, 197
Milner, S. R., 173, 178
Mineralogy, 79, 105, 145
Minerals, 105, 207
Mitchill, Dr., *Theory of Contagion*, 99, 101
Mitscherlich, E., 80ff., 85–6, 144–5, 147
Molecules, 76–7, 86, 146, 150, 175, 181, 183, 216; molecular weights, 79, 83ff., 146, 186f., 189, 202, 205, 218
Monge, G., 43, 54, 93
Moniteur, 117

Monnet, A. G., geologist, 21
Moore, T. S., 225
Moray (or Murray), Sir R., P.R.S., 4
Morley, E. W., 45
Morris, Gouverneur, *Diary of the French Revolution*, 23–4
Moseley, H. G. F., 177, 227–8
Mulder, G. J., *and* physiological chemistry, 148; *and* 'protein', 152

Nagel, D. H., 227f.
Nash, L. K., *and* atomic theory, 66
Natural History Museum, 125
Nature, 217
Neale, Sir P., 4
Nernst, W., 176, 178
Newman, Mr., *and* Royal Institution, 116, 153
Newtonian atom, 67, 72, 87; Newtonian-Dalton atom, 86
Nicholson, William, *Dictionary of Practical and Theoretical Chemistry*, 94; *Journal of Natural Philosophy*, 64, 71, 77, 96, 102, 139
Nitre, 32, 34ff.
Nitro-compounds, 202
Nitrogen, 45, 64, 99, 113, 118, 203.
  See also atomic theory *under* Dalton
Nitrogen chloride, 115
Nitrogen dioxide, 12
Nitrous ether, 32
Nitrous oxide, 7f., 11, 66, 99–102
Nomenclature, chemical, 45–6, 78–9, 135, 142, 168–9, 188; pharmaceutical, 142
Notes and Records, 168
Nuclei, theory of, 200
Nuovo Cimento, 84, 186

Octane, 208
Odling, William, 199, 210, 214f., 217, quoted, 83–4; *and* Karlsruhe conference, 190f.; molecular weight, 146, 205
Oefele, A. von, *and* sulfine and sulfone compounds, 213

Index

Oersted, H. C., 127, 158
Ogston, A. G., 228
Ohm, G. S., 174; Law, 174f.
Onsager, L., 180f.
Organic analysis, 32, 51, bodies, 43, 45
Organic chemistry, 47f., 79f., 85, 104, 137, 142, 149f., 185, 187, 191, 198–215, 229f.
Organic compounds, 201ff., 207ff., 211–13, 215–16; organo-metallic compounds, 209
Ostwald, W., 176
Oxford:
Junior Scientific Club, 83
School of Chemistry, 223–32
See also Balliol; Christ Church; Jesus; Keble; Lincoln; Magdalen; Queen's; Trinity; and under Laboratories
University Museum, 226, 228f., 231
Oxides, 28–9, 67–70, 73, 141; metallic oxides, 77, 143, 145
Oxygen, 5, 9, 11–12, 29f., 33, 37, 40ff., 44f., 47–9, 64, 66, 75, 79f., 93, 95f., 99–101, 107–10, 113, 132, 141, 201; and electrical changes, 177
See also atomic theory under Dalton

Parcieux, de, and Paris water supply, 25
Paris, Dr., and Davy, 101, 112
Particles, 64–7, 69–72, 75, 161, 174, 186
Pasteur, L., 85f., 135, 188
Paulze, Jacques, Lavoisier's father-in-law, 23, 54
Pavesi, P., 84, 185, 192
Payne, Mr., and Royal Institution, 116, 153
Pebal, L. von and Gerhardt, 214
Peel, Sir Robert, and Davy, 125f.
Pepys, Samuel, with reference to Royal Society experiments, 4
Peroxides, 226

Petit, A. T., see Dulong
Philosophical Magazine, 104
Philosophical Transactions, 61, 104, 124, 168
Philpot, J. St. L., 228
Phlogiston, 6–7, 9, 11, 27, 33, 36f., 39, 41, 45f., 48, 92, 107, 109, 111, 135
Phosphorus, 26f., 29f., 45, 81, 115, 144, 196
Photo-chemistry, 13, 73, 228, 231
Physical chemistry, 12, 153n., 158, 172, 227, 229, 231
Physiological chemistry, 48–51, 99ff., 138, 148. See also Animals; Humans; Metabolism; Respiration
Pickering, Spencer, 227
Piria, R., 186
Planck, M., 176f.
Platinum, 124, 148, 225; vessels, 139, 143; wires, 159f.
Playfair, Sir Lyon, 207, 209
Pneumatic chemistry, 1, 3, 5–7, 12, 99–102
Pneumatic Institute, Bristol, 96f., 100f.
Poisson, S. D., equation, 178
Polymers, 81, 147; polymeric, 148, 152
Poole, T., friend of Davy, 130
Potash, 106, 108f.
Potassium, 106–9, 113
Povey, Thomas, F.R.S., 4
Priestley, Joseph:
Biographical information and characteristics, 1, 8f., 12, 14–18, 58; experiments, 2–15, 17; and oxygen, 33–4
Other references, 19f., 23, 28, 31f., 36, 44, 55, 136
Protein, 148, 152
Proust, L. J., 59
Prout, William, and hydrogen, 75
Prussian Academy Berichte, 86

Quantum theory, 177
Queen's College, Oxford, 223, 230

242 Index

Raffles, Sir Thomas Stamford, 126
Raikes, Humphrey, 228
Ramsay, Sir William, 5
Raoult, F. M., 176
Ray, John, 125
Rayleigh, John William Strutt, third Baron, 5
Regnault, H. V., 187
Respiration, and transpiration, 27, 36f., 40f., 45, 48–50, 99
Richards, Theodore, 143
Richter, A. G., 88
Richter, J. B., 138–9
Rivett, A. C. D., 225
Roberts-Austen, Sir W. C., 227
Robinson, Elihu, meteorologist and instrument maker, 59
Rolleston, Sir Humphry, 93
Romsey Public Kitchen, 103
Roscoe, Sir H. E., 188, 190, 226
Rose, G. and H., 59, 138, 144; and atomic weights, 80
Ross, Professor Sydney, 168
Rothpletz, F. A, and geology, 197
Rouelle, chemist, and Lavoisier, 20
Royal Institution, the, 69f., 74, 102–6, 111, 114ff., 119, 132, 154ff., 168, 213
Royal Society, the, 4, 20, 43, 58, 74, 87, 106, 110, 120, 122, 127, 129, 154, 163–6, 210; Presidency, 124–5; Royal medals, 127
Rozier's *Journal*, 34
Rumford, Sir Benjamin Thompson, Count von, 63, 95, 102ff.
Ruskin, John, 197–8, 226
Russell, A. S., 230
Rutherford, Lord, 87, 177, 227

Sack, H., and conductivities at high frequencies, 180
Safety lamp, 119–23. See also under Davy
Sage, B., and phosphorus, 26
Salts, 77, 81, 111, 141, 176f., 181f., 225; chromium salts, 178; subacid and superacid salts, 74–5, 77, 139

Sanderson of Oundle, 222
Saussure, N. T. de, 72
Savich, V. I., and Karlsruhe conference, 188, 190
Scheele, C. W., 19, 58, 80, 134f.
Schischkoff, L. M., and Karlsruhe conference, 188
Schmitt, R., and succinic acid, 213
Scott, Sir Walter, with reference to Lady Davy, 114; to Davy's *Salmonia*, 130
Seguin, M., 23, 49f.
Shelburne, Sir William Petty, first Marquis of, and Priestley, 8f., 14, 17
Sidgwick, N. V., 183, 225, 228f.
Silicon fluoride, 12
Smith, Henry, 225f.
Society of Medicine, 48
Soddy, F., 227f.
Söderbaum, Dr., biographer of Berzelius, 135
Sodium, 106–8
Solutions, aqueous, 174ff., 183, dilute, 175ff., 181. *See also* Conductivity
Somervell, D., 225
Southey, Robert, and Davy, 97–9; and 'laughing gas', 9, 101
Stahl, G. E., 6, 27, 41, 141
Stas, J. S., 143
Stenhouse, J., chemist to the Mint, 215
Stephenson, George, and safety lamp, 121
Stereo-isomerism, 85
Stock, A., and Karlsruhe conference, 188, 192
Stodart, J., surgical instrument maker, 155–6
Stoichiomtry, 138
Stokes, Sir G. G., 210
Stokes's law, 179–80
Story-Maskelyne, M. H. N., 225
Strecker, A. F. L., and Karlsruhe conference, 188, 191
Strontium, 108
Structural chemistry, *see under* Kolbe

Index

Sulphur, 27, 45, 81, 115; sulphates, 46; sulphites, 46; sulphur dioxide, 9
Sutherland, W., 177
Swedish Royal Academy of Sciences, 134f., 146
Synthetic chemistry, 213

Tellurium, 109
Tennant, Smithson, 118, 122
Thackray, A. W., with reference to Dalton's atomic theory, 66, 68, 71, 89
Thenard, L.-J., 107–10, 136
Thermochemistry, 39–42, 230
Thermodynamics, 176
Thomson, T., 59, 65, 67, 70, 75, 136; *System of Chemistry*, 70–1, 74, 89; *History of Chemistry*, 71
Tizard, H. T., 225
Tobin, Dr., friend of Davy, 93, 131
Torricellian vacuum, 64
Travers, M. W., 5
Trinity College, Oxford, 223, 226–9
Trudaine, D., and Lavoisier, 33
Tyndall, J., 208, with reference to Faraday, 172

Ulich, H., 182

Valency, 83, 179, 181, 209f., 215f.
Vapour densities, 81, 83, 145, 186f., 189
Vauquelin, L. N., 100
Vegetation, and air, 5–6
Veley, V. H., 224, 229
Ventilation, 49f., 156
Volhard, J., and sarcosine, 213
Volta, A., 102, 138
Voltaic pile, 109, 170
Voskresensky, A., and Mendeleev, 193

Walden, A. F., 182, 230
Walton pyrites, 7
Washburn, E. W., 182
Water, 25–6, 42–5, 72, 136, 167; anhydrides, 205; conductivity of, 175; formulae for, 80, 84, 113, 142, 186, 205, 218; soda water, 3; temperatures, 61, 70. *See also* aqueous *under* Solutions
Watt, Gregory, and Davy, 97
Watt, James, 14, 99n.
Wedgwood, Josiah, 14
Wedgwood, Thomas, 97
Weltzien, K., 84, 187f., 192, 218
Wheatstone, Sir C., 155, 174
Whewell, W., 168f.
Whistler, Dr. Daniel, 4
Wiedemann, G. H., 174
Wien, W., 180
Williamson, Alexander, 185, 204ff., 210, 215, 217; and etherification, 146, 205; Gerhardt's *Comptes Rendus*, 82; Gerhardt's types, 213; molecular formulae, 83f.
Wilsdon, B. H., 225
Wilson, D. R., 224
Wilson, George, *Religio Chemici*, with reference to Dalton, 66
Winmill, T. F., 225
Wöhler, F., 132, 136, 138, 147, 151, 185, 188, 206; and atomic weights, 80; *Dictionary of Chemistry*, 210; and organic chemistry, 199
Wolfenden, J. J., 228
Wollaston, W. H., 59, 74, 77, 80, 122, 124, 127, 130, 136, 139, 157
Wordsworth, William, and Davy, 98, 104 n.
Wurtz, C. A., 84, 185, 187f., 190ff., 218, and primary amines, 204
Wynne-Jones, W. F. K., 228

X-ray data, 179, diffraction, 86, spectra, 228

Young, Arthur, *Travels in France and Italy*, quoted, 23, 44
Yule, C. J. F., 224

Zeitschrift für Electrochemie, 182
Zinc methyl, 208–9, 212
Zittel, K. A. von, 197
Zoological Society, London, 126, 131

Randall Library - UNCW   NXWW
QD15 .H37
Hartley / Studies in the history of chemistry.
3049001385943